I0605242

The Cynic's Guide to **Wine**

Published in 2025 by Académie du Vin Library Ltd

academieduvinlibrary.com

A CIP catalogue record for this book is available from the British Library

ISBN 978-1-917084-63-5

Photos: Pages 35, 99, 219 courtesy of Daniel Ogulewicz; Page 4 by Cup of Couple via Pexels; Page 10, Steve Race/Stockimo/Alamy Stock Photo; Page 36, iStock.com/halans; Page 65, iStock.com/Armastas; Page 66, iStock.com/Sicha69; Page 100, iStock.com/mspoint; Page 131, iStock.com/Esperanza33; Page 132 © Mick Rock/Cephas; Page 159, iStock.com/jumaydesigns; Page 160 © Jean-Bernard Nadeau/Cephas; Page 192, iStock.com/gilaxia.

Publisher: Hermione Ireland
Editor: Rebecca Clare
Design: James Pople
Index: Catherine Hall

Printed and bound in Latvia, by PNB Print

The Cynic's Guide to **Wine**

Sunny Hodge

Contents

Introduction

The first sparks that ignited a hospitality flame within me came towards the tail end of a Mechanical Engineering degree at University College London. Most graduates were scooped up and placed into financial institutions for their newly acquired skills of taking complicated systems apart, analysing them, and then putting them back together in a more fruitful manner. We spent years poring over textbooks on thermodynamics, materials chemistry, fluid dynamics and triple integrations knowing we may never need to use what we learned in our future careers. Engineers develop a particular methodology for learning and understanding, and apply this to the world around them.

My choice to earn a living in wine and hospitality was neither expected nor planned. At every level of an operating restaurant is a sense of organized chaos; the pace, the noise and smells, the accents and raucous laughter all had me intoxicated – and although my scientifically inclined mind would hate to admit it, it was love at first sight. Whilst studying, to put food on the table I worked at a nearby Portuguese chain restaurant called Nando's, a casual dining spot which excelled in grilled peri-peri chicken. It was in casual dining where I learned the most about people and friendship; our restaurant was indiscriminatingly open in its hiring policy and took on board any vaguely approachable person with a slight interest.

English speaking wasn't always a requisite, and previous restaurant experience wasn't needed either. This hiring strategy meant that the training process needed to be rigorous, and as I moved up the ranks into a managerial role, learning how to deliver concepts without always having a common language was essential in keeping the restaurant gears turning.

A number of restaurant experiences later I found myself working for the likes of the great Paulo de Tarso and Nicholas Jaouën to open one of London's most desirable and service-focused Italian restaurants, Margot. Think white tablecloths, Damien Hirst originals, Hollywood movie stars, and tailored dinner suits for the waiting team. From humble chicken grilling beginnings, I found myself hosting industry legends at their peak. To be tested and tasted by Michel Roux Jr and Massimo Bottura had the kitchen team in tears of ecstasy; there was a strong sense of 'we finally made it' in the air. I came on board as a Duty Manager and both Nick and Paulo guided me masterfully on how to lead service, conducting maître d's, head waiters, bartenders and sommeliers like an orchestra. Service was an art, all in sync, in place and melodious. We would discuss drink and food with guests for hours on end, and these were guests who had high expectations and knew what they liked. Like most, I learned to sell wine through vivid tales of vineyards, winemakers, appellations and the expectations of varietal character – just as an actor would commit lines to memory. I discovered on the front line of service that the guests had also developed a similar narrative around wine. Food and cocktails were quantifiable and spoken about technically: ways of curing, time in lemon juice, millilitres of vermouth, it all made logical sense. But wine was shrouded in mystery. Sommeliers spoke of clay and chalky soils but couldn't explain what difference it made to the wine. High-spending regulars asked for Chablis whilst making it clear how much they abhorred Chardonnay, there were Sancerre lovers who detested Sauvignon Blanc, and an overwhelming majority of old schoolers who vehemently despised organic wines without having the foggiest idea what 'organic' really meant in the context of wine. Wine catchphrases and teachings had been passed down from one generation to the next, by word of mouth, like some ancient religion spread orally and set free to flow far and wide.

The first ever wine books I bought were for my own development in Margot, fully aware that after a decade in hospitality wine was still my weakness. I just couldn't get my head around it, so I picked up a few volumes covering the wine regions we sold most: Italy and Bordeaux. The books gave wines scores out of ten, told me what flavours I should taste and what sorts of soils made up these prestigious wine regions. Later in service, a bottle of Sassicaia was decanted, worth a typical team member's monthly wage. I heard the sommeliers speak of how the gravel there is comparable to Medoc, which gives 'courage and stony minerality' to the wine. I questioned if these descriptors made any sense, and asked if indeed these connections were true. How could I possibly feel comfortable regurgitating to guests what I had learned from experienced sommeliers without knowing in myself if it was fact or not? In reading those books I hoped to fill the wine-knowledge void within. However, they left me with more questions than answers. I followed up with wine courses and still found myself back at that same place. I was fed more lines to reel off to guests, learned the 'what' instead of the 'why'. I eventually came to realize that something is fundamentally broken with the way we have all come to understand wine. That moment of clarity awoke the engineer in me and this scientific adventure began.

By 2018 I had just about saved up enough to open Diogenes the Dog in Elephant and Castle. It is a neighbourhood wine bar, the primary focus of which is to aid guests in demystifying wine. Just as in my peri-peri chicken days, teaching ideas without a common language has been crucial to the success of the concept. We removed the old language of wine by taking all possible familiarity away from the wine list, importing wines from as far afield as Texas, Taiwan and Poland to create a list of total unknowns which made the teaching and reprogramming of wine understanding that little bit easier. The list cut away all the pretentiousness, assumption and ego which often accompanies wines we think we know. The concept was a risk, my first enterprise alone, and something that had never been attempted before – a lifetime of savings all thrown into one basket.

After years of the blood, sweat and tears that are an unavoidable part of the success of any new business, Diogenes grew to be a hit in the London wine scene, winning multiple awards, and had even given me the opportunity to further our teaching potential across another venue. Aspen & Meursault opened in leafy Battersea just a few years later and took the same approach across an array of low-intervention wines. These last eight years have been a personal journey to decode wine, removing the mysticism and finding the answers that our guests and wine professionals have all been looking for, yet don't dare to ask. A true understanding of wine spans many fields of science, and to present the information in this book I've also had to take many paths. The little floor space available in my London flat is forever littered with research papers on viticulture, oenology and taste perception. I've scoured universities across the UK and attended lectures on subjects ranging from soil science to microbiology. Perhaps the greatest impact has come through speaking directly with the winemakers, regenerative farmers and leading oenologists that we work with to further support these findings. Nowadays I write, judge, mentor and consult on wine. It's a world filled with experts, journalists, buyers, Masters of Wine and Master Sommeliers and it is clearer to me now than ever that there remains a divide between wine assumption and sound scientific understanding. In this book I rip away the marketing and tasting notes and present you with the scientific links from soil to glass that will enable you to understand how it all works. You'll learn the basic and elusive 'why', the scientific foundation we all need whether you are a seasoned drinker, or like myself, simply have a keen interest in understanding how all things work. This book is the start of a revolution in the way we understand, speak about and celebrate wine. It's the book I needed when starting my journey, and one that I hope will serve you well on yours.

"Something is fundamentally broken with the way we have all come to understand wine."

1.
Roots

Roots

Those of you who are reading this to get a practical understanding of wine are in for a treat. This guide is one of the quickest tools for getting your head around this tricky and often romanticized subject. In the sciences we initiate the study of each subject from first principles. We have followed this same process in food since the early 1900s, beginning the teaching of food science to chefs with technical subjects such as the Maillard reaction, so let us begin this process with wine. Our truth-finding wine journey starts underground.

Experts are always telling us that soil influences flavours in wine, coffee and all sorts of other food made from plants. Yet how many of us truly understand how soil makes a difference? We hear this sort of dogma regularly in wine settings: 'This wine tastes the way it does because of the soil' or 'That wine smells like this because of the "terroir".' Before checking the validity of these statements in reference to the soil we need to understand how roots function. This is the key to separating the more 'dreamed up' aspects of wine marketing and sales from the scientific truth of what makes a difference to our glass of wine. Developing an understanding of what vine roots can and can't do, and how these resource scouts interact with their surroundings are the first principles at play here.

There are two mind-bogglingly crafty mechanisms that vines have evolved in order to capture nutrients from the soil. Since these leafy climbers are firmly grounded they need to harness the powers of chemical magnetic attraction to get a hold of the vital stuff all plants require to stay alive. The first of the processes they employ is 'cation exchange'. As the second most valued reaction in nature, behind photosynthesis, it is vital to maintaining the health of plants.

Cation exchange

This process fascinates those working across the full span of the food chain, from Michelin-starred chefs to farmers. Cation exchange provides the real 'aha' moment when it comes to figuring out how plants interact with the soil. It's the first step to gaining a better understanding of wine (it will help you look after your houseplants better too) and it will change the way you think about plant life.

Cation exchange is utilized by all plants to trade nutrients and minerals with the soil. Plants shoot out their roots to infiltrate the earth, and these roots facilitate the swapping of minerals with the soil around them. We've covered quite a lot of ground in that one last sentence, so perhaps we should pause to touch on some key terms before moving forwards. The 'minerals' we mentioned are a bit of a buzzword in wine in relation to both soil and taste, but precisely what are they and do they determine the flavour of wine like so many experts claim?

What are minerals?

Minerals are made up of organized and symmetrical arrangements of ions bonded together. An ion is simply a single atom or group of atoms with a net positive or negative charge. It's this net charge which enables them to attract other ions, like a magnet.

A single three-dimensional arrangement of ions is referred to as a lattice. If this lattice pattern grows and repeats itself in a regular fashion, it is in turn called a crystal or a crystalline structure: these are what we mean when we talk about 'minerals'. Crystals come in many forms and most are not sparkly or gem-like, instead they are dull, weirdly shaped lumps. To sum up, a crystal is made up of an ionic lattice structure, repeated. This crystal is referred to as a 'mineral' in soil, viticulture and farming parlances.

On each grubby outer surface of minerals, some charge polarity exists (i.e. a negative or positive charge). This net charge allows minerals to attract other similar ions, hence the mineral will steadily grow in size over time.

If a mineral surface is unobstructed and in the open, it will grow to form the pretty, crystalline looking structures we're more familiar with: smooth, elegant, faceted shapes. These unobstructed conditions in nature are not common, and most crystalline structures grow in small, tight spaces. Minerals shoulder to shoulder with their surroundings are also looking to grow, and when hemmed in by their neighbours, they grow into whatever strange irregular shapes their environment permits. This more common growth scenario results in minerals looking like the grubby lumps we often see in the soil.

Minerals are naturally occurring elements or compounds of elements, and it's rare for them to exist independently as one single element. Common minerals we may have heard of in a wine context include those of quartz in the Llicorella soils of Priorat, chalk in Champagne or basalt in Etna. Minerals are said to be a key factor in the unique characteristics of the wines from each of these distinguished wine regions. We will look deeper into types of rocks and minerals in the next chapter to see if this is indeed true. Soils are often referred to as having a high mineral content, usually in reference to vine health and fertility, yet this can only make technical sense if referring to one specific mineral in relation to another, since all soils are largely made up of minerals. Terms like sand, silt and clay often crop up when taking tours of vineyards; all are examples of different-sized mineral structures.

While we are on the subject of minerals, you are probably wondering 'What's a rock, then?' A rock is a combination of different sorts of minerals smashed together in various ways. The boundaries of the mix of minerals in a rock do not share electrons in a crystalline structure as mineral compounds do, instead they're bonded in numerous different ways, all of which we'll cover in more detail shortly.

When I started my studies in wine, I found all these rock- and soil-related terms appeared muddled together. To this day soil terms are used to explain how wine has come about but this terminology is frequently and confusingly used inaccurately. A clear understanding of the components of soils will aid us in dismissing the wine mystics. In simple terms soil is a mixture of minerals, gases, liquids, living organisms and organic matter ('humus' for wine buffs and soil geeks).

"Soil is a mixture of minerals, gases, liquids, living organisms and organic matter."

Now that we have a clear understanding of the basic definitions of minerals and soil, let's go back to our cation exchange. As we've learned, rocks (which are made up of minerals) in the soil rarely have complete electrical neutrality on their outer surfaces. This exterior electrical charge is the same charge which allows them to attract similar particles and grow. This charge also attracts a range of other charged ions present in the soil. A positively charged ion is referred to as a cation, whereas negatively charged ions are known as anions. These cations and anions are present in the soil water, and as this water moves through the soil, the positive cations form loose bonds with the negatively charged mineral surfaces they meet on the way (as with magnets, opposites attract in an effort to gain charge neutrality).

We're left now with our rocks and minerals in the soil coated in a thin layer of these nutrient cations. I use the term 'nutrient' because of the essential health benefit to plants these cations provide. They are still, in the simplest terms, just positively charged ions. Their charge attractions to mineral surfaces are weak because of the assortment of differing charges they carry and the random collection of solids dissolved in the water. In all likelihood these ions will not be able to make a perfect, super-strong charge fit on their first attempt. Yet a small step towards some charge neutrality is better than none at all. Once this weak bond is formed, a cation will stay put on the surface of the mineral until one with a stronger charge comes along to displace it. A hierarchy is formed where the strongest charged cations dominate for prime position on the negatively charged mineral surface, displacing the weaker ones in the process. This ionic dance on mineral and rock surfaces is eternal, a wonderful and chance chemical exchange where cations get thrown off by ions with a stronger charge and are washed away by the soil water (until they find a new negatively charged surface to attach to) in a never-ending cycle of attachment.

In addition to the strength of the positive charge, the size of these passer-by ions also makes a huge difference to the charge hierarchy. Smaller ions take priority since they're able to be physically closer to the mineral surface and can therefore form much stronger bonds. An easy way to picture this charge relationship is by visualizing two magnets of opposing polarity across the table from one another. The stronger the power of one magnet, the larger the force of attraction across both. The physically closer they are to one another, the exponentially greater the force of attraction between them will be.

The technical term for that weaker bond which connects an ion to a mineral is 'adsorption', as opposed to 'absorption' which would mean it became part of the crystalline structure. In farming and viticulture the measure of the total negative charges within the soil that can adsorb plant nutrient cations is called the cation exchange capacity (CEC). The higher the CEC, the more nutrient cations are available in the soils. This is a direct measure of soil fertility.

From a viticultural perspective, grapevines don't ask for much and can flourish in some shockingly poor soils. They are true survivors in the plant world, as are many other climbing vines. Instead of growing resource-heavy structural components such as tree trunks and branches, they use external structures and other plants to gain height and capture sunlight. With limited resources they can achieve a lot.

The top-dog cation at the pinnacle of the cation exchange charge hierarchy is hydrogen (H^+). Hydrogen ions are a single hydrogen atom which has lost an electron, making the hydrogen ion just a single proton in size, one ten-thousandth the size of your average atom or ion. Consequently, this tiny cation can exert an extremely strong positive charge and always takes prime position on mineral surfaces, cuckooing its way onto rock surfaces and displacing many lesser nutrient cations in the process.

Now for the big reveal. The beauty in all of this, the real feat of nature and evolutionary development, is that grapevines have inbuilt hydrogen cation factories which can metabolize and create these highly attractive ions on demand. Vines draw

---

"From a viticultural perspective, grapevines don't ask for much and can flourish in some shockingly poor soils. They are true survivors in the plant world."

hydrogen (H) from water (H_2O) via their roots. Photosynthesis allows the water molecule to be separated into oxygen (O), which is released into the atmosphere, and hydrogen ions, which are then pumped out into the soil via the roots to facilitate cation exchange.

These hydrogen ions then go to work displacing hordes of weaker nutrient cations found on neighbouring minerals and other soil components, freeing them from mineral surfaces momentarily before nearby roots scoop them up and into the vine. An understanding of this process is key to ascertaining what plants can and can't take up. Whatever our soils are made up of, it's the tiny nutrient cations in the soil water and on rock surfaces that are taken up by plants, not the minerals or rocks themselves, which simply act like magnets to store and hold nutrients. This is a key principle for developing our understanding of wine, and already we can begin to dispel myths of a wine incorporating specific sorts of minerals and rocks from the soil, since they are simply too big for the roots to take in directly.

Grapevines have a good level of control over what nutrients their roots take up. Transpiration is the process of water evaporating from plant leaves, creating a pressure surge that draws water up through the vine. The pressure created through transpiration is what allows plants to suck water up from the roots. Newly exchanged and nutritious soil water is then forced through a layer called the Casparian strip, which sheathes the cylindrical outer surface of the root. This waxy strip contains materials such as suberin and lignin, which make it impervious to water penetration, ensuring the flow of mineral water is kept in the central section of the root. Here the nutrients are thrust through rigid cell walls, and into root endodermal cells which act as gatekeepers for water and nutrient uptake. The vine takes what it needs and discards the rest.

This guarded control system is not entirely foolproof, and the vine can sometimes get itself in a muddle. This can especially occur when there's an abundance of similar looking nutrient cations in the soil, which leads to deficiencies in other doppelgänger nutrients. An example of this occurring in the field is vine iron deficiencies, a common problem in limestone-rich soils. Iron is involved in the synthesis of green chlorophyll, therefore deficiencies have a major effect on the photosynthesizing power of the plant and can lead to chlorosis or yellowing of the leaves. We will explore the intricacies of this phenomenon further on (see p. 27), so you'll have to take my word for it, just for now.

In a nutshell, that's cation exchange, the process of how all plants electromagnetically 'swap' nutrients in the soil.

Rhizosphere interactions

A rhizosphere is the region of soil around the roots where the chemistry and microbiology are altered as a direct result of the roots. It is not a region of definable size or shape, but consists of an ever-reducing gradient in biological, chemical and physical soil properties.

The root–soil interactions within this space illustrate the glorious connectivity of nature. For centuries we were unaware of the mechanisms involved, but these days, at least in the geekiest of regenerative soil literature, it's all anybody is talking about. Rhizosphere interactions work to benefit the vine's nutrient uptake and can assist in many ways that geology cannot.

Vine roots exert a tremendous amount of pressure through their tips to grow and carve their way through the soil. To assist in movement during this growth, the root cells secrete mucilage. This grim-sounding liquid is viscous, dense and insoluble and acts to reduce friction whilst the root is growing. Beyond lubrication, mucilage also provides physical protection, and assists the roots in nutrient acquisition. Whilst doing this it also quite impressively binds soil particles together, massively improving soil quality by creating tunnel-like structures which aid in water infiltration and the aeration of soil. As if mucilage hasn't already done enough, amongst other uses, it also serves as 'bait' for root pathogens and works to sequester toxic soil metals. Mucilage is simply what plant dreams are made of.

As well as the gift of mucilage, all plants produce compounds aptly referred to as rhizodeposits, which are secreted into the soils from their roots. Root rhizodeposits include organic acids, amino acids, proteins, sugars, phenolics and other secondary metabolites which are generally fed to micro-organisms. And, since I know you're wondering, secondary metabolites are ingredients that don't assist directly in primary metabolism, such as growth or reproduction, but can assist in other indirect ways such as plant defence. The type of chemical cocktail released by roots is heavily shaped by the specific plant species and climatic conditions. The chemical rhizodeposit in turn then goes on to shape the make-up of the microbial community within its rhizosphere, which in its turn is instrumental in shaping the local plant life. One of the many examples demonstrating how roots can easily sway the rhizosphere environment is by the secretion of acids, which spurs the release of certain nutrient cations through localized acid erosion and will also alter specific nutrient availabilities in the soil. We could spend hours looking at the many ways roots interact with soils but, at risk of losing you, I'll just touch on the following two.

Bacterial symbiosis – assisting with nitrogen

Within the rhizosphere there is a plethora of symbiotic bacterial relationships which can come to the aid of plants. Legumes are a great example of this. In most organic forms, nitrogen compounds are too large to be taken up by roots. When legumes can't get enough nitrogen from the surrounding soil their roots exude chemical signals in the form of flavonoids, which beckon the invasion of certain rhizobium bacteria. These little bacterial helpers march right in and set up camp in the legume roots, forming little nodules that look like tiny bird houses. Here the bacteria can live in safety and begin work on turning large nitrogen compounds (N_2) in the soil, into ammonia gas (NH_3), which another neighbouring bacterium then breaks down into the soluble nitrate (NO_3^-), a form of nitrogen which local plants can take up into their roots. This symbiotic process is the main mechanism utilized in nature to provide nitrogen to plants. Though nitrogen is the most abundant element in the air we breathe, plants can't take it in directly from the atmosphere. This problem is the driving force behind our extensive use of nitrogen-based fertilizers in agriculture today.

Lupines are very common legumes that were widely grown in vineyards throughout the Roman era. Philosophers and writers of the time, such as Theophrastus and Columella, recommended the use of lupines and other legumes to assist in vine growth but it is only very recently that agricultural research has uncovered the science behind what the Romans knew millennia ago.

Another very useful bacterium that lives in the soil, and doesn't require the housing that legumes can offer, is *Azotobacter*. This bacterium also works to benefit plants by fixing atmospheric nitrogen and making it available to plants. There have been many studies on *Azotobacter* which illustrate that they prefer neutral to alkaline soils. Even marginally acidic soils, of just pH 6, show little to no presence of these nitrogen powerhouses. When we get into matters of soil, we will learn which rocks affect the acidity of soils and therefore influence the amount of natural nitrogen available in the soil.

Mycorrhizal symbiosis – for all the other nutrients that the plant and animal kingdom can't provide

As a curious child I was forever lifting up rocks to find out what was hidden under them. It was the creeping things that caught my attention then, and I paid little attention to the wiry root-like structures that wove their way through the soil. I now know that these were mycorrhizae, a term derived from the Greek, meaning 'fungus root'. If you are a gardener you may have noticed this furry, beardlike structure at the end of plant roots. Mycorrhizal fungi are thinner than plant roots, accessing areas that plant roots cannot, and they end up growing into all sorts of places. By latching onto the roots, they create humongous nutrient and communication exchange networks in the rhizosphere – almost like an internet in the soil.

There are two main types of these mycorrhizal associations with plant roots: the ones that we see above ground fruiting as mushrooms (ectomycorrhizae) and the hidden types that only live under the soil (arbuscular mycorrhizae). Both associations serve to increase the contact area between plant and fungi, extend nutrient trade routes and enable the exchange of hard-to-find nutrients which the plant requires, for carbon (C), which the fungi need. Plants have access to plenty of carbon from photosynthesis (taking in CO_2 from the atmosphere), but fungi can't photosynthesize. This valuable mycorrhizal exchange provides the fungi with carbon, which is tricky for them to source in other ways. The fungi in return provide their partner plant with additional water via the extended mycorrhizal network they have created, plus access to nutrients that vines would not normally be able to get hold of such as phosphorus (P), iron (Fe) and micronutrients like zinc (Zn) and copper (Cu). The variability in nutrient availability in soils is a complex subject but these four are especially tough to get hold of, and each for their individual reasons. Phosphorus for example, reacts quickly with minerals in the soil, forming strong compounds with iron, aluminium and calcium. This locks up the phosphorus, preventing plants from accessing this crucial nutrient on their own. Phosphate-solubilizing fungi can unlock this nutrient and get their due rewards from sharing it with a plant.

These sorts of mycorrhizal associations are evidenced to increase the tolerances of plants to drought, as well as reducing salinity stress and oxidative stresses, and increasing pathogen resistance, all while improving soil structure in a similar manner to roots. Yes, more mucilage!

Mycorrhizal associations

Ectomycorrhiza

These only make up around three per cent of mycorrhizae and occur mainly around the roots of woody plants like trees. They form a dense fungal sheath over the tree's root tip, throwing a net of fungal filament around the root cortex cells, but they do not penetrate the root cell walls. These fungi form fruiting bodies and, mind-blowingly, it's these three-percenters which make up all the visible mushrooms we see above ground.

Arbuscular mycorrhizae

These sorts of mycorrhizal roots do penetrate the root cells and form a highly branched structure. They're the most abundant of all mycorrhizal associations and are completely dependent on their partner plant for their carbon intake. These mycorrhizal networks don't produce mushroom-like fruiting bodies, instead they produce their spores inside swollen structures called 'vesicles', located along their root networks. These are emptied into the soil when the conditions are just right, to be dispersed by soil water onwards to other locations. Arbuscular mycorrhizae also help plants deal with overloads of toxic heavy metals in the soil, such as arsenic, cadmium, chromium, copper, lead and zinc, which can often lead to stunted plant growth. They do this either by preventing these heavy metals from getting through their own mycorrhizal network, shielding intake into the vine roots, or by increasing the plant's tolerance to these heavy metals steadily over time. Their not-so-toxic relationship is a purely dependant symbiosis where if the mother plant dies, so will the arbuscular fungi, so the fungi are incentivized to do a good job in keeping their partner plants alive.

Just how many nutrient ions do these vines need?

For the plant lovers and green-fingered hobbyists among you, these next few pages may provide insight into the health of your houseplants and garden as well as broadening your understanding of wine. The grape species used for most wine production is called *Vitis vinifera*. It doesn't require a lot in the way of nutrients – yet balance is the key. The risk to a plant's metabolism is not governed by the quantity of nutrients it has available, but by the one nutrient it has the least of. Any one nutrient deficiency is enough to weaken the entire plant; therefore the success of a vine's health is dependent on its access to that lesser available nutrient. We should bear in mind also, that as the plant grows bigger, so does its nutrient requirement. Thus, as one nutrient deficiency may be plugged artificially by human intervention, another may occur as the plant grows in both size and nutritional requirements.

To give you a better understanding of what makes each nutrient so special, and its key function in plant health, the next few pages give a brief rundown of the 13 principal plant nutrients.

"The risk to a plant's metabolism is not governed by the quantity of nutrients it has available, but by the one nutrient it has the least of."

Principal soil nutrients

Nitrogen

What it's needed for: a core component in metabolizing chlorophyll, building amino acids for growth, and creating nucleic acids, hormones and compounds like pyrazines, etc.

Ideal concentration: 500–1,200 parts per million in soil, depending on vine age and yield potential.

Symptoms if there's not enough: older leaves become pale green or yellow, growth is stunted, young shoots become pink or red, and in severe cases necrosis (rapid cell death) occurs.

Where it's found: predominantly made available by nitrate fixing soil bacteria, just small portions are derived from nitrates in the air caused by decomposition of organic waste.

Phosphorus

What it's needed for: phosphorus is essential for plant growth. It is a component of cell membranes and DNA, and plays a vital role in photosynthesis, the movement of sugars, and carbohydrate storage within the vine.

Ideal concentration: 20–50 parts per million in soil.

Symptoms if there's not enough: the reproductive processes of the vine are generally affected by phosphorus deficiency before vegetative growth is affected. Affected plants see reduced fruit setting and fertility, resulting in a reduced yield.

Where it's found: rhizobium soil bacteria release, humus decay.

Potassium

What it's needed for: potassium is required for a range of essential plant functions. In photosynthesis, potassium regulates the opening and closing of stomata (pores in the leaves that allow gas and water to be exchanged between the plant and the air), therefore regulating carbon dioxide uptake. Potassium also triggers the co-enzymes required for energy production (ATP), and plays a fundamental role in water regulation.
Ideal concentration: 75–100 parts per million in soil. Too much can lead to a decline in calcium and magnesium uptake in the vines due to these three cations having similar chemical properties. This negative effect on the uptake of other nutrients is referred to as nutrient antagonism.
Symptoms if there's not enough: chlorosis, the yellowing of leaves due to chlorophyll deficiencies, or in extreme cases the death of plant tissue (necrosis) can occur.
Symptoms of excess: when vines are grown in soils with extremely high levels of potassium (possibly due to contamination or agrochemicals), we often find higher quantities in the final wine, although this very rarely exceeds 0.15 per cent of the wine. Potassium found in grape juice will have a direct influence on the final acidity of the wine. It is one of very few instances where the specific nutrient make-up of the soil so directly influences our wine. Note, however, that here we mention nutrient make-up of the soil, with no reference to geology.
Where it's found: cation exchange with humus and clays derived from potassium feldspar or muscovite, and animal urine.

Sulphur

What it's needed for: sulphur plays a critical role in amino acid formation, required for growth, development and hormone responses. Like nitrogen, sulphur is needed for the construction of many amino acid chains, which are the building blocks of proteins. Sulphur is also critical in the process of photosynthesis, and an essential nutrient involved in the biological nitrogen fixing process carried out by soil bacteria.
Ideal concentration: 10–200 parts per million in soil.

Symptoms if there's not enough: sulphur deficiencies are often misdiagnosed as nitrogen deficiencies since both help the vine retain its green colour. Yellowing leaves, associated with sulphur deficiencies, show up on newer leaves. In contrast, nitrogen-deficient plants show this symptom on older leaves. The reason for this is that nitrogen is easily transported in plants and has the fluidity to move from one location to another. Sulphur, on the other hand, does not move through plant tissues as easily.
Where it's found: many different types of rhizobium soil bacteria can aid in the release of sulphur.

Calcium

What it's needed for: involved in the development of enzymes, starch, cellulose, cell walls, and cell signalling. Optimum calcium leads to improved berry quality, reduction of grape stalk necrosis, improved berry firmness and greater storage potential as a result of stronger structural membranes in the grape.
Ideal concentration: 50–2,000 parts per million in soil. Has nutrient antagonism with magnesium and potassium.
Symptoms if there's not enough: low calcium concentration causes browning of the grapes and eventually leads to them drying out.
Where it's found: cation exchange with humus and clays derived from calcium feldspar or calcite.

Magnesium

What it's needed for: magnesium has several functions in the vine. It is the central component of chlorophyll. Magnesium also serves as an enzyme activator for several carbohydrate metabolism reactions. In addition, the element has both structural and regulatory roles in protein synthesis.
Ideal concentration: 100–250 parts per million in soil. Has nutrient antagonism with calcium and potassium.

Symptoms if there's not enough: most often chlorosis or necrosis.
Where it's found: cation exchange with humus and clays derived from mafic materials or dolomite.

Iron

What it's needed for: iron is an essential component of many proteins and enzymes and plays an important role in chlorophyll formation, photosynthesis and respiration.
Ideal concentration: approximately 20 parts per million in soil
Symptoms if there's not enough: chlorosis or reduced yields.
Where it's found: iron is the second most abundant material in the Earth's crust and is widespread in rocks. However, it is a bizarrely common occurrence for vines to suffer from an iron deficiency, and the chemical environment needs to be just right for the roots to take up iron. Iron can only be taken up by roots in its soluble (Fe^{2+}) form. In alkaline soils, iron is most likely to be stable in its insoluble (Fe^{3+}) state, which explains the well-documented chlorosis issues that often occur in limestone soils. In addition to this, the soluble form of iron can be leached away easily precisely because of its soluble nature, causing a deficit of usable iron. In its soluble state it is attained via cation exchange with humus and clays derived from mafic materials or iron oxides.

Zinc

What it's needed for: zinc is involved in the synthesis of plant growth substances (phytohormones), is a catalyst involved in cell metabolic reactions, and plays a role in pollination and fruit set.
Ideal concentration: approximately 2 parts per million in soil
Symptoms if there's not enough: mottling of young leaves (patchy discolouration), necrosis, irregularly shaped, almost sharp, leaves, uneven growth spurts.
Where it's found: cation exchange with humus and clays.

Manganese

What it's needed for: manganese plays an important role in chlorophyll synthesis and is an enzyme co-factor for metabolic reactions.

Ideal concentration: approximately 20 parts per million in soil.

Symptoms if there's not enough: banded chlorosis on leaves; severe deficiency can affect vegetative and reproductive growth and may delay berry development.

Where it's found: cation exchange with humus and clays.

Copper

What it's needed for: copper is a component in the creation of enzymes of oxidation, in chlorophyll synthesis and in the formation of lignin during cane ripening.

Ideal concentration: approximately 0.5–10 parts per million in soil.

Symptoms if there's not enough: poor vine growth and yield, short canes with shortened internodes, small, pale-coloured leaves.

Where it's found: cation exchange with humus and clays.

Molybdenum

What it's needed for: amino acid building and enzyme functions.

Ideal concentration: approximately 0.0004 parts per million in soil. Molybdenum is a trace element found in soils and not common. In addition to this, only 2–20 per cent of molybdenum found in soils is in a form that the vine can take up, hence even with such small requirements, deficiencies can occur.

Symptoms if there's not enough: poor vine growth, very small leaves, leaf cupping and edge burn, papery feel to leaves, rubbery feel to shoots, wood failing to develop, poor fruit set.

Where it's found: anion exchange with humus and clays. Anion exchange is where anions are adsorbed by the plant roots by physical contact, a flow of mass via water intake, or diffusion of anions from a high concentration to a lower concentration area. Levels of anions in the soil are measured using the anion exchange capacity (AEC), and the AEC of soils is always lower than the CEC. The AEC, however, is directly affected by the acidity of soils, since higher acidities tend to increase the AEC.

Boron

What it's needed for: boron plays a key role in the internal regulation of growth by plant hormones, pollen tube growth and metabolism. High levels of boron can cause phytotoxicity (in other words, they are poisonous to the plant).
Ideal concentration: approximately 0.02–12 parts per million in soil.
Symptoms if there's not enough: boron deficiency can reduce crop yields because of poor fruit set, cause chlorosis and lead to a high proportion of seedless berries, which appear flattened.
Where it's found: anion exchange with humus and clays.

Chlorine

What it's needed for: chlorine is used in enzyme activation, in osmotic balance, and is essential to the chemical reaction that allows the opening and closing of the plant's stomata, making it instrumental in photosynthesis.
Ideal concentration: N/A
Symptoms if there's not enough: chlorine deficiencies are rare as chlorine is abundant on the surface of the soil; however, when they do occur, deficiency symptoms include wilting due to restricted and highly branched root systems, and leaf mottling.
Where it's found: mostly on the soil surface; chlorine is also dissolved in rainwater or sea spray.

Enzymes 101

We have talked quite a bit about enzymes, but what exactly are they? Enzymes are biological catalysts, they exist in all living things and almost always as a protein. They work to speed up the rate of a specific chemical reaction in a cell by lowering the activation energy required for a reaction. They are not destroyed during a reaction and can be used over and over again. I think of them like biological power tools. They require energy to work, they aren't alive hence can't act alone, and they speed up the job whilst being utilized many times till they eventually break!

A single living cell contains thousands of different types of enzymes, each specific to a particular chemical reaction. Their names usually have an '-ase' suffix, like lactase or protease. In the realm of food and drink, enzymes have always played a key role and continue to do so even in the age of industrial-scale food production. We couldn't produce lactose-free milk without them, as it's the enzyme lactase which cleaves the sugar lactose into two smaller, simpler sugars, making it much easier to digest for those suffering from lactose intolerance. The purpose of the protease enzyme is to break down proteins. Found naturally in raw tropical fruit like pineapples, papayas and mangoes, it is used in many food scenarios, from tenderizing meats to removing bitter proteins in cheese.

Enzymes are also responsible for speeding up the browning of food and wine when exposed to oxygen. Using a grape as an example, when the fruit splits and the cells in the tissues break open, they release polyphenol oxidase which utilizes oxygen in the air to create a brown pigment. Many chefs add lemon to cut fruit to slow this process down, since the citric acid denatures the polyphenol oxidase enzymes, and stops them from functioning. Enzymes will come up frequently in this book, especially in later chapters when we explore what makes up the smell and taste of wine.

How much water is enough?

Like most worthwhile things in life, some struggle is required. When photosynthesizing, the vine can choose what to do with the energy it produces. A happy vine, in a good place with lots of water, sunshine and nutrient-rich soil, will not want to move. It's doing well, and because of this will want to stay put and continue to grow. It'll want to grow fast and spread itself out to take as much sunlight for itself as possible, gaining a competitive advantage over its neighbours. To do this, the vine must concentrate its energy on green vegetative growth. The fruit from these happy vines won't be very concentrated since the efforts of the vine are channelled into making more leaves, tendrils and shoots as opposed to better fruit. This green growth strategy is referred to as 'vigor' in wine circles.

Let's take the same vine, significantly reduce the water, and place it on the verge of dying. For its lineage to survive, it'll go on what it believes to be a suicide mission. The vine musters its energy towards the furthering of its offspring, the grape, in the hope that its genes will be selected and passed onwards to a kinder environment. The brighter, more flavoursome and pungent the fruit, the more likely it is to be spotted, eaten and transported. This vine may well die in time, but knows that it has sacrificed all to give a fighting opportunity to the future of its genes.

It is the latter of these two scenarios that makes better wine, and the simplest way we can ensure a vine adopts this survival strategy is to restrict the amount of water it has. This is easily done in drier climates where water intake can be controlled through irrigation, ensuring the vine neither becomes too vigorous nor weakens too much, either of which would be to the detriment of fruit development. In wetter and more unpredictable climates it is harder to control vine watering. This is a large factor in understanding how vintages can affect the cost and quality of wines so much from one year to the next. Amongst other environmental factors, very rainy years yield reduced quality fruit, leading to lower value vintages, if indeed the wine gets released at all.

Roots are constantly on the lookout for water and nutrients, so if there's not enough water in the upper layers of topsoil, they will dive down deeper to find more reliable supplies. Longer roots are a certain bonus for the attainment of groundwater. In drier spots where water access may be tough, or in locations where rainfall is inconsistent, deeper roots may be the difference between life and death for our vine. Big root networks toughen the vine against dehydration; though it's worth noting that nutrients are far less plentiful the deeper down in the soil we go.

Rootstock

Towards the end of the eighteenth century, global trade was at an all-time high, fuelled by the slave trade and other trade routes created off the back of it. Travel between the New World of the Americas and Europe was easier than ever – no customs, no restrictions – but this unrestricted movement created problems, especially for European vineyards.

The mid-1850s saw the arrival of a little louse called *Phylloxera vastatrix*, fresh off the boat from North America – it destroyed up to ninety per cent of Europe's vineyards. This little aphid-like creature loved eating the roots of our favourite winemaking grape species, *Vitis vinifera*. A devastating pest with a staggeringly complex 18-stage life cycle, it can even adapt to forming wings in moist conditions. Phylloxera is a nightmare for vine growers, and at its most dangerous when in the nymph stage of development, devouring roots and injecting poison into the vine. This weakens plants and leaves open wounds at the roots, putting them at risk of further bacterial or fungal infections. These infections invariably finish the job.

In the middle of the nineteenth century, Europe was the epicentre of wine, and European vineyards were in tatters. A 20,000 franc reward (equivalent to almost £1 million today) was issued by the French Minster of Agriculture and Commerce to find a 'cure' for phylloxera. To this day, no cure per se has been found and phylloxera still lives amongst us in European soils.

However, a nifty work around was found, and it came about from further investigating the source of the issue. North America had numerous local American grape species that lived alongside phylloxera and still thrived. On closer analysis, Europeans found that these American vine species secreted a thicker, stickier sap which clogged up the mouths of the pesky phylloxera nymphs whilst creating a sappy protective layer over the wounds, thus preventing further infection. However, most of these indigenous American vines were said not to make good wines.

Nowadays in most of Europe, *Vitis vinifera* vines are individually grafted to these American vine roots to overcome the phylloxera plague. A swift and effective resolution: Eurasian *Vitis vinifera* drawing water and nutrients through American rootstock. The roots are protected against phylloxera but the grapes are still the same varieties that produce good wines.

By the time this apt solution was discovered, vine growers and wine producers in Europe were at the very brink of collapse. They needed to get back into full flow, and luckily, it was back to business again. The effects of this solution were little researched since growers and producers were just happy to be trading. Since Europeans had a monopoly on wine, it meant they could maintain their lead at the helm of the ship without having to change their wine traditions or wine species for post-phylloxera wine production.

Even now we don't give the practice of grafting much thought or weight – it is simply something that has to be done. Our wine market is still oriented around varietals of *Vitis vinifera*. We choose Pinot Noirs, Malbecs or Chardonnays but give little thought to the sorts of rootstocks they are grown on. Will we ever see a world where rootstock like Rupestris du Lot, 1103 Paulsen or Kober 5BB is listed on a bottle of Picpoul di Pinet or Pinot Grigio from our local wine shop?

"We don't give the practice of grafting much thought or weight – we choose Pinot Noirs, Malbecs or Chardonnays but give little thought to the sorts of rootstocks they are grown on."

Grafting

The practice of grafting can be traced back more than 4,000 years to ancient Mesopotamia and China, and it makes use of the lack of an immune system in plants. This means different varieties or even species of plants can grow from one another. With animals, the immune response would detect the foreign biological material and attack on sight (which is why people who have had organ transplants have to take immunosuppressants).

Rootstock research in the field of apple growing has shown conclusively that different rootstocks do lead to taller or shorter apple trees, and have a measurable effect on fruit development. Viticulturists know all too well that there are drawbacks from the graft: it's a weak point, and will always be a wound. Grafts can lead to issues in sap flow and are most susceptible to frost damage or rot.

Many American rootstocks are hybrids of two different vine species, created to make best use of both parent characteristics. Rootstocks can be selected for pest resistance, drought tolerance, soil acidity regulation, and even to manage vigour. *Vitis rupestris*, for example, is a highly vigorous American species. Grafting *Vitis vinifera* to a rootstock from this species will give better resistance to high lime content in the soil, and the roots will naturally draw in far more water and nutrients at a faster rate than the original *Vitis vinifera* roots would. It's like putting an articulated truck engine inside a Ford Fiesta. If such major changes can be brought about by roosstock choice, why aren't we talking about this more? 'Terroir' is regularly mentioned in the context of wine but should we not be talking about rootstock too?

2. Debunking 'terroir'

Debunking 'terroir'

Soils and 'terroir' are frequently referred to on wine labels and at wine tastings. We are told that 'This stunning 2014 Cabernet Franc has grown from granitic soils, which accounts for its freshness and structure'. We listen and nod, and some agree. I've always questioned statements such as this, it's not clear-cut, and it's hardly ever challenged. How does the soil make any difference to the wine we drink? And why do we rarely pose this question to the experts?

This issue reminds me of a Uruguayan trade tasting I attended some years ago. The small room was packed full of wine journalists, buyers and wine-trade personalities, all huddled around the table on a Zoom call to the winemaker. The glass in question was a Malbec. The wine writer to my right took her first sip and asked what soil the vines were grown in. The winemaker responded authoritatively, 'Some sand, gravel and calcareous soil', to which the wine writer erupted, 'What!? No clay, how can this be?' Taking another sip, she spat with precision into our communal spittoon, and remained indignant. I had an urge to jump in and ask what had upset her so much, but social etiquette got the better of me.

The winemaker reiterated that there is *almost* no clay, but there may be a little in some parts of the vineyard. Our wine writer looked at her audience triumphantly before frantically jotting down some crucial notes, and the tasting moved on.

What was it she tasted? How could she link what she tasted to clay, or was it all just a performance? And why don't we put this much soil emphasis on other food or drink? Surely root vegetables such as potatoes and carrots should be more affected, as they develop surrounded by soil.

Throughout my years studying wine through traditional channels and time spent at the front end of hospitality I was unable to find a definitive answer, nor could I find anybody in the wine world who could provide me with any scientific reasoning behind their assertion of the link between soil and flavour. Further research was required.

Even before humans farmed, they had a mystical reverence for land and soil. It is the source of all plant life: tiny seeds grow into plants which in turn go on to sustain human and animal life. Without the understanding we have now of photosynthesis, respiration and the importance of water, our ancestors naturally credited the soil with that supernatural power. Perhaps when we look to geology and soils to explain what we taste in wine we are harking back to such ancestral beliefs. Does our millennia-long relationship with the soil cause us to place too much emphasis on its influence? Science has now demonstrated that plants can be grown without soil. Hydroponics shows us that gorgeous fruit can be grown with just water, light, air and trace elements.

In this chapter we dig deep into soil science to find if there is a link between soil and what we taste in wine. Much of my research has been self-guided and outside of the traditional wine learning channels. Soil, as we've learned, is made up of a mixture of minerals, living organisms, humus, liquids and gas. This mixture of aggregates acts as a conduit for the supply of water and nutrient ions. We'll now look at each of these constituents to better understand how they contribute to the supply of water or nutrient ions to our vines.

Minerals

Almost all geological rocks are tasteless and odourless. This being the case, when wine buffs state they can taste the clay, chalk or slatey minerality in wines grown on certain soils, what on earth are they all talking about? (No pun intended.)

As we learned in the previous chapter, it's not these larger complex compounds and structures which we often talk about in wine that are taken up by the roots, but the surrounding nutrient cations which happen to adsorb onto their surfaces. If we popped a calcium-rich bone into our bowl of cornflakes in the morning, we wouldn't expect it to do much more to our calcium intake unless we actually ate the thing. Yet we often make assumptions in wine that soils rich in certain minerals will directly add that mineral's associated qualities to the vines or even to the wine we drink. It is important to keep in mind that soil minerals are not the same as the nutrient cations which surround them. If someone suffers from anaemia, let's say, the doctor will offer the iron nutrient in the form of a supplement as a remedy, not some rusted iron nuts and bolts.

"Does our millennia-long relationship with the soil cause us to place too much emphasis on its influence?"

Surely there's a way we can get these minerals broken down and into the vines?

Physical weathering is one way that rocks both above and below the soil can be broken down into smaller constituents. Continual elemental bombardment will wear a rock down. In addition to this, daily expansion and contraction of bedrock due to fluctuating temperatures from day to night will cause movement and therefore erosion. And in really cold regions, constant ice expansion and contraction in crevices will also whittle down these rocks into smaller pieces.

Chemical weathering is a part of life. Minerals, like our bodies, are constantly being degraded by the environment. Chemical weathering by acids, oxygen, salts and even water changes the molecular structure of rocks. These processes can lead to discolouration and are responsible for some iconic soils such as the famous terra rossa soils of Barossa Valley, where surface iron oxidized to form a rusty reddish-brown combination of hematite, limonite and goethite.

Rainwater is usually also a little acidic because it has atmospheric carbon dioxide dissolved in it. The action of rainwater degrades rock both above and below soils to form a chemical solution that will eventually affect the chemical make-up of the soils. Both chemical and physical weathering provide a steady source of cation nutrients to soils, and as one can imagine, the effects of weathering above the soil are a little more aggressive than below. Logically, then, the make-up of soil nutrients due to weathering has more to do with minerals from rocks and other materials above ground, transported via rainwater, than those below. Which makes you wonder why so much importance is placed on the underlying geology of vineyards.

We mentioned earlier that *almost* all geological rocks are generally odourless and tasteless – the exceptions to this rule are halide minerals such as sodium chloride (salt). Salt can rapidly dissolve and provide us the taste of salinity on our tongue. Salting soil ensures that nothing will grow there, since plants will do their best to reject large sodium intake. This prevents plants losing their own water content via osmosis of fluids from a low internal sodium concentration to a high external sodium environment. For this reason, most wines contain well below the perceivable threshold of salt, and any salinity we do perceive is almost certainly figurative.

While I am sceptical about the perception of mineral elements in wine, it's hard to believe that we conjured aroma associations with certain metals and rocks out of nowhere. I can myself recall the most distinct smell of lead from an unfortunate turn of events in Peckham, London. In my early twenties I ended up with a couple of lead air rifle bullets lodged in the back of my head and jaw when I was the victim of a clichéd London robbery following a skate filming session (yes, I'm an avid inline skater) that was quickly dampened by rain. After the shooting I scrambled to get away onto the first double decker bus that would stop for me, while bystanders looked on aghast. I didn't realize how seriously I had been injured until the adrenaline subsided. To this day, the smell and taste of that alien object under my skin is unforgettable. How is it possible that smell didn't exist?

In reality, the smells we perceive to be attributed to metals and rocks are in fact related to tertiary factors and chemical reactions rather than the object itself. Coins and iron-based metals are a great example since we all know they have some smell when handled. In the case of coins, the familiar ferrous smell is due to the oils and sweat from our skin reacting with the iron. The same smell occurs with blood when rubbed between your fingers: it exudes an identical metallic smell due to the iron content reacting with our natural skin oils. A newly minted, untouched coin has no smell.

Rock faults, cracks and joints

Water sources can be influenced by both the physical shape and chemical make-up of rocks. Rock displacements can often change the direction and pattern of groundwater passage in the bedrock, either guiding water to the vine or drawing it away. An open mouth of a weaker part of the rock that has given way is often picked apart by erosion and widens with time. This fault zone is then filled by lighter and softer sediment which settles in the widening displacement. Roots can manipulate these softer zones and break through to find valuable potential sources of water.

Joints are even smaller fractures in the rock. These can also be picked wider apart by erosion and can often get concreted, filled by underground waters which leave precipitate minerals behind in the gaps to form veins – frequently the minerals left behind are calcite and quartz and bear no relation to the parent rock.

When vine roots hit bedrock, they ordinarily cannot pass through. However, the tiny hairs which surround the roots will seek out water, and if they sense water in a small fracture, they'll divert root growth in that direction. In extremely dry conditions, these little oases of water in fractures and faults are pivotal in keeping the vines alive.

Soil colour

Before the bars, when life was a little more spontaneous and carefree, I'd take weeks out of work to get on my motorbike, an ancient Triumph Bonneville, and tour through Europe. It was the open road, a few books, my skates and me, with little planned outside of tomorrow. Looking back, I was totally underequipped for the journeys, in an era where mobile data was pricey and out-of-country GPS apps weren't that reliable. My natural sense of direction is practically non-existent, so having the attitude that getting lost was part of the journey was a must. I'd drive till late, and as darkness drew in, finding a bed for the night became the next priority. Cheap hostels or inns were ideal, but if nothing was near and it was dry, camping rough was often an option. On colder nights in Spain, the black of the tarmac under a bridge or off a side road was a good find. It was warm to the touch, and if I was tired enough, as comforting as any high tog duvet. Seeking out the darker coloured spots kept me warm through the cooler parts of the night – at the time I didn't think much more of it.

The colour of our soils is defined by its mineral content, water composition and the type of organic matter present. Soil colour is measured using a scale called the Munsell colour system. Soil colour definitively does make a difference to the progress of vines and plays a much greater part in defining terroir than we often give it credit for.

Colour influences the rate of soil warming in the spring and cooling in autumn. Dark soils absorb more heat than more reflective light-coloured soils. In addition, soils with higher moisture content – being darker – absorb more solar radiation and will warm up or cool down much more slowly than drier soils do. Water has almost twice the heat capacity of dry rock but holds a lower thermal conductivity. Hence the moisture content of soils will increase the overall potential capacity for heating up, and will maintain that heat over a longer period of time. Many growers wet their soils before impending frost to allow the soils to take in more heat during the day, and release it slowly at night, thus avoiding frost settling at night. Clay soils in the vineyard also contribute to heat retention because they have a higher heat capacity than sandy soils, primarily due to their unique water retention abilities.

Soil temperatures define the microbial health of soils; warmer soils will promote microbial life. Heat also plays a role in increasing root metabolism and nutrient uptake, increasing the rate of starch breakdown and hence metabolism in the roots. This encourages root growth, shoot growth and an earlier budburst for our vines. The manner in which a soil retains heat is influenced by its thermal conductivity, and it's worth observing that thermal swings are far less prevalent the deeper one delves down. In many cases it is the colour and water content of vineyard soils that are responsible for avoiding catastrophes such as frost on late budding grape varieties. Soil colour and water content, regardless of soil type, play a huge part in grape development and metabolic rate.

"Soil colour definitively does make a difference to the progress of vines and plays a much greater part in defining terroir than we often give it credit for."

Reflectivity also makes a difference, and the colour of soil has a direct impact on the amount of sunlight reflected back up photosynthetically into the vine canopy. This can influence grape yield and the development of sugars, anthocyanins and polyphenols via increased photosynthesis. The proportion of reflected photosynthetic light is measured at the vineyard on the albedo scale. An infinitely dark soil that absorbs all light would have 0 per cent albedo and if it was possible for all light to be reflected then the soil would have 100 per cent albedo. Real-life vineyard soils range from around 35 per cent for lighter calcareous soils, and just 10 per cent for darker basalt soils.

We can also conclude from this that if sunlight is reflected upwards, then a large proportion of heat is also reflected away from our soils, leading to cooler soils. Conversely, darker soils reflect less light but absorb more heat. This means our soil colour permits our vines to benefit from either the additional heat from absorption of sunlight or from extra photosynthetic light from reflection, but we won't reap both benefits in one location.

Types of vineyard rock

Only by understanding how vineyard rocks are formed can we fathom if they make a difference to our vines and wines. I have listened to and read so much about the different rock and soil types throughout my wine career, but sadly much of it was quite jumbled up. I have tried to organize and simplify the science here since I assume that most readers have a limited interest in rocks. There are three main categories of rock: sedimentary, igneous and metamorphic.

Sedimentary rocks

As the name implies, sedimentary rocks are formed from weathered rock sediments deposited by wind and water that bind together in numerous ways to one day form a solid mass. Over the years small, weathered rock particles are dispersed by wind or water and erode the other rocks they skitter across on their journey. These particles settle on the ground or the seabed and in time other sediments settle above them. The pressure of the world above soon gives these sediments the energy to form strong bonds and they become one united rock. If this process is associated with river sediment, we refer to it as alluvial. If the small particles find their way into a sea, we call the sediment marine. These adjectives are regularly thrown around in wine talk, so it is worth knowing what they mean. One of the Polish wines we import from Winnica Wieliczka comes from vines farmed on glaciomarine sedimentary soils. That sounds like rather a complicated mouthful so let's break it down. The city of Wieliczka is famous for sitting above one of the largest salt mines in Europe. Prior to the invention of refrigeration the region amassed great wealth by selling the salt for use in food preservation. The huge salt deposits were all that remained from an ancient ice age ocean. The soils inherited by Agnieszka and Piotr, the power couple behind these biodynamic wines, are termed glaciomarine in reference to the old frozen, then melted, now dried up ocean that once existed where their vines grow. Just like the soils of Chablis, they are littered with fossils, including the woolly mammoth tooth which is represented on all of their wine labels.

We can further classify sedimentary rocks by particle size, explained below from largest to smallest.

Boulders

These large, angular rocks most often form near the source of original bedrock since they're too big and heavy to travel far.

Cobbles, pebbles and gravel

All three of these provide excellent drainage, with cobbles being the largest, pebbles a touch smaller and gravel the smallest. The famous 'galets' of Châteauneuf-du-Pape AOC are a type of cobble. Like many larger rocks in vineyards with large diurnal temperature ranges (the difference between night and day temperatures) they can assist in the ripening of the grapes by absorbing the heat from the daytime sunshine, then releasing that heat steadily overnight to keep the vines metabolizing. This may not always be a good thing as, like us, vines do need their rest, thus it is a climate- and vineyard-specific benefit.

Sand

Sand in soils is, as one would guess, of a similar size and coarseness to the sand found on beaches. A sand's colour is determined by the type of bedrock it is originally weathered from. The black basalt sands of Hawaii and the purple quartz–manganese sands of Pfeiffer Beach in California are two examples that remind us that sand is a reference to particle size and not material. As a side note, phylloxera nymphs are not a huge fan of sandy soils as the grains are loose and create an uncomfortable travelling environment. Because of this there are still a few sandy pockets of ungrafted vines that have survived in Europe from the era prior to phylloxera.

Silt

These rock fragments are just about visible with a magnifying glass. They have a floury texture when dry, and are easily transported by wind or water. When moved from one place to another by wind, we refer to them collectively as 'loess'. Over 10 per cent of the world's surface is covered by loess. Both silt and loess have great heat retention properties and decent water retention due to the smaller particle size and absorbency. They can also be a bit of a nuisance for vineyard irrigation systems, with small particles causing blockages down the line.

Clay

The finest of all sedimentary rocks are referred to as clays. The use of clay in this context just refers to particle size, although rather confusingly clay is also used as the name of a material, and this is the use with which we're all more familiar. And yes, clay is a clay! Clay particles are so tiny that even hand microscopes can't see them, but their small size gives them an extraordinary surface-area to volume ratio. If we think back to cation exchange, we can envisage how having a humongous surface area will directly increase a material's CEC. Hence, clays have disproportionately high nutrient storing properties, increasing the CEC and fertility of soils which contain them.

When clays get wet, they gain plasticity and can be spun into many shapes. Wet clay doesn't permit much water through, increasing the water retention properties of clay soils. Too much clay in wetter conditions may pose an issue – if wetted, clays form a watertight seal that can result in the waterlogging of vines. Inversely, when dry, clays are susceptible to hardening and cracking, which can also be problematic.

The word 'loam' refers to soils which are a balance of the three smallest particle sizes: clay, silt and sand.

How clay shapes wine

The versatility and unique properties of clay do go some way to explaining our fascination with this soil type in relation to wine. It is therefore worth looking a little deeper into some of the different types.

Smectite is a form of clay from the montmorillonite group with an extremely high CEC. The particles can adsorb a lot of water molecules, leading to significant expansion when wet, hence they're often dubbed *swelling clays*. In extreme weather conditions these clays can be problematic. When very wet they can expand so much they effectively block up the soil, preventing water drainage and causing water to build up above the clay layer, leading to waterlogged roots.

When very dry, they can shrink and crack so much that a soil can become too aerated. Air is not a good medium for cation exchange, so these large, dry air gaps in soil directly reduce the vine's ability to access nutrients. The shrink–swell properties of smectite are also potentially disastrous for surrounding building foundations, and it is a common cause of subsidence in London properties. The montmorillonite group of clays can also be referred to as bentonite, which is often used as a fining agent in wineries.

Kaolinite and illite. In contrast to smectite, these clays expand and contract very little. Kaolinite has a much larger particle size than illite at five times its size, which, for perspective, is more than 250 times the size of smectite. While the CECs of both are significantly lower than that of smectite, they do provide better drainage. Kaolinite is often used for pottery and became known as 'China' after its discovery in south-east China.

Vermiculite. Within its crystal structure, vermiculite can permit its silicon to be replaced by both aluminium and magnesium. This creates quite a sizeable charge imbalance, and results in an exceptionally high CEC, higher even than that of smectite. Vermiculite contains water, like most clays, but keeps that water bonded wholly within its structure, which is highly unusual. If heated to the point where the water boils, the release of steam causes the crystal lattices to expand. These extreme temperature conditions aren't relevant in the vineyard, however expanded vermiculite is utilized in vine nurseries as it's easy to use, lightweight, drains well and maintains a high CEC. Vermiculite clays have further commercial applications in bricks, the brake linings of cars and cat litter.

"When very wet, clays can block up the soil, preventing water drainage, leading to waterlogged roots."

Limestone

These sedimentary rocks often steal the limelight in vineyard narratives due to their unique properties and are formed under very specific conditions. Rainwater is slightly acidic; it slowly dissolves rocks and washes away as a solution. This solution ends up settling somewhere and drying out (for example, in shallow seas or on cave floors) until the dissolved components solidify and form deposits. We refer to these deposits as evaporites. Rocks formed in this manner, with high proportions of calcite evaporites, are called limestones. Unlike clays, limestones have great water drainage properties. Their physical composition allows water to pass through even when they are saturated, enabling roots to continue working at peak performance with little fear of waterlogging.

Travertine is a limestone that forms when calcium carbonate is deposited in freshwater environments, particularly groundwater lakes and hot springs, where rapid precipitation can occur. These deposits are typically large, dense and banded. When the deposits are highly porous, so that they have a spongelike texture, they are typically described as tufa. Tufa is abundant across large swathes of Italy, Hungary and many US states. It is not to be confused with the igneous rock *tuff*.

Sea water contains high levels of calcite and aragonite, which can precipitate to form a calcareous mud on the sea floor, in the same way that orange juice or soup will separate into their constituent densities over time. If this mud hardens, we have another limestone. This underwater calcareous sludge is often made up of the remnants of dead animals and microscopic organisms. Chalk is formed in this way and is very relevant to vineyards; it's a limestone derived from the compression of microscopic plankton skeletons that settled on the sea floor millions of years ago. Chalk is synonymous with the soils of Champagne and southern England.

The main component of all types of limestone is calcium carbonate, which carries a pH level of 9.9, making limestone alkaline rather than acidic. We can now conclude that the more limestone a soil contains the more it will tend towards pH 9.9, thus increasing the overall pH of the soil and lowering its acidity. By the end of this chapter we will have a clear idea of what effect soil acidity has on a vine.

Fossils

Fossils also fall into the class of sedimentary rocks and are a hot topic in wine regions all around the world. These are formed from the harder parts of animals or plants, such as leaves, wood, bones, teeth and shells, that are rapidly covered in sediment. Picture a decaying velociraptor carcass swept by the sands of time. The sediment forms a protective layer which stops further decay, pressures will increase from other sediment building up above our velociraptor, and water may also seep in. Now add a healthy dollop of time (say, 10,000 years) and the sediment becomes 'lithified' – turned to stone. Lithification is like cementing, where water, time and pressure create strong bonds. We're left with a perfect rock mould of our long-deceased velociraptor. It's often assumed that wines take on the taste of the fossilized remains found in their soils. But these fossils are just rock casts and can in no way make our wine taste like the creature from which they originated. We're often led to believe that we can taste the ocean in a wine or told how well it goes with seafood as a result of the Kimmeridgian soils of Chablis being formed from fossilized sea creatures. It may well pair wonderfully, but not because of the fossils.

During a recent Chablis Masterclass run for the wine trade by a truly knowledgeable Master of Wine (MW), I couldn't help but enquire: 'What is it that specifically changes the profile of wine grown on the much-idolized fossilized Kimmeridgian and Portlandian soils of Chablis compared to any other old limestone soil?' The MW's response was: 'It's the make-up of the soil.' I asked again for specifics. At this point the room sat silent, until the MW said, 'It's the uniqueness of these soils which changes the wine.' I felt then my initial question was comprehended, but an element of manoeuvring from the MW was required to dodge this bullet. I couldn't let this one go, so I asked for a third time what specifically differentiates the Kimmeridgian soils from other limestone soils, and sadly I got the final brush off: 'It's

---

"Fossils are just rock casts and can in no way make our wine taste like the creature from which they originated."

the properties of the soil, perhaps water retention.' The MW moved on to the next question from the audience, leaving my question with no real answer. A single expert or scientist cannot have all the answers, and we should not shy away from admitting this when speaking about wine, a field of expertise where so many hide behind the mysteries of geology.

Igneous rocks

These are rocks that were once molten, heated below the earth's surface and then exposed to the open air through gradual erosion or violent volcanic eruption. Notable igneous rocks include granite, obsidian and pumice.

Granite

Found across the globe, predominantly near plate boundaries, granites form when magma is cooled many kilometres below the Earth's surface. As the volcanic mountains above the magma are eroded away, the molten rock is slowly lifted to the surface and solidifies. Granites are rich in quartz and contain mostly potassium feldspars. Granites tend to be dark, influencing the overall colour properties of the soil they sit in, and as we well know this in turn affects the state of vines grown in that soil. Granite is a common vineyard rock and notably makes up the soils of Beaujolais, Northern Rhône and Priorat.

Obsidian

Obsidian is formed through the quick quenching of magma, often cooled rapidly by sea or lake water. The hyper-cooling gives insufficient time for any crystal structures to form, hence obsidian is a naturally shiny, non-crystalline, black rock. The presence of obsidian causes extreme darkening of soils and therefore heavily influences their thermal properties. As you can perhaps gauge from the complicated formation process, obsidian in soils is quite rare, however it can be found in parts of California and Oregon.

Pumice

This light-coloured, extremely porous rock formed as a result of violently explosive volcanic eruptions. The rock's large pores are created by trapped gas bubbles made during the rapid cooling of a frothy magma. Some pumice rocks are so porous that they can float on water, until their pores slowly fill up with water and they sink. This high porosity means they have great water retention properties, and a high CEC due to all their holes and the increased surface area within. In addition, the large pores are perfect breeding grounds for microbial life and smaller soil organisms, which have a positive effect on the long-term fertility of pumice-based soils. I'm frequently asked in service for volcanic wines but when I ask why, nobody can quite put their finger on why they want or like these wines. And although we know how pumice changes soils we are still a step or two away from understanding how it may or may not change our wine. Pumice is very common; examples are found in vineyards across Santorini and large parts of the USA.

Metamorphic rocks

These begin life as a rock (sedimentary, igneous or other metamorphic). As the name implies, they go through a change from their original form when under high heat, high pressure, exposure to hot, mineral-rich fluids or all three. Conditions like these are most common deep within the Earth or where tectonic plates meet. Some great metamorphic vineyard rocks include slate, marble and serpentine.

Slate

Slate is a fine-grained, metamorphic rock formed by further compression of sedimentary shale, mudstone or basalt. It has extremely low water retention properties, to the point of making it 'waterproof', and holds great thermal and chemical stability. It's often remarked that slate-rich vineyards produce slatey and mineral characteristics in wine but as we know now, this cannot be the case. The colour of slate ranges from light greyish purple to darker shades of greyish black, depending on the carbon content of the slate. Slate is synonymous with some pretty epic wine regions such as Corsica, Mosel in Germany and Banyuls and Corbières in France.

Marble

This type of rock is created from the metamorphism of sedimentary carbonate rocks, usually limestone or dolomite. Those derived from limestone will alter the acidity of the soil. Many well-known wine regions are adjacent to marble quarries, including parts of Burgundy and Tuscany.

Serpentine

This is derived from the metamorphism of magnesium-rich olivine (a green magnesium–iron silicate that looks like kryptonite). Most plants can't tolerate this rock due to the overly high levels of magnesium in serpentine-rich soils. The high levels of magnesium lead to a reduced uptake of calcium and potassium nutrients by nearby plant life, a well-documented phenomenon where a surplus of one nutrient causes a decline of another, called 'cationic nutrient antagonism'. (This was mentioned in the last chapter when we summarized the nutrient properties of magnesium, calcium and potassium.) One theory in explaining this problematic interaction is that these three similar looking nutrients compete for the same nutrient transport pathways in plant roots, so a surplus of one leads to a dilution of the others when lined up along the same single pathway. Additionally, serpentine contributes an element of toxicity to soil in the form of increased nickel, cobalt, iron and mercury levels. Like kryptonite, you really don't want a lot of it around.

"Geology certainly is responsible for many effects on a vine's development, but does not affect wine in the way that wine writing would lead us to believe."

We're beginning to build a far more accurate picture of what geology does and doesn't do to wine. It certainly is responsible for many effects on a vine's development, but does not affect the wine in the way that wine writing would lead us to believe. In summary, geology does have an impact on the colour, drainage, thermal properties, nutrient availability and sometimes the acidity of soils, but we can't determine whether these are good or bad for vines until we look at other vineyard factors. For example, do we want dark soils in a hot climate? What variety of grape are we planting? There are far more impactful variables to work with than geology and, as we will soon learn, many ways that farming can completely bypass the effects of geology. When you also consider the fact that all soils are aggregates of many different types of rock, under constantly changing weather conditions, it seems like a lot to take in. Is this why, when selling wine, we rely on the romanticism of geology rather than the litany of other variables?

Micro-organisms

In modern sciences we often think of micro-organisms as the engines of life. The first living colonizers of this planet were bacteria, algae and lichens, and all three played a part in accelerating the process of weathering to form today's soils.

Lichens can grow where not much else can, like rock surfaces. A lichen is not a single organism, but a stable symbiotic association between a fungus and photosynthesizing algae and/or cyanobacteria. Like fungi, lichens require carbon as a food source, kindly provided by their symbiotic photosynthesizing algal partner. Mosses also play a big part in the weathering of rocks, releasing those nutritious cations back into the soil.

The simple beginnings of soil came together 250 million years ago, when these first rock colonizers came into existence. Soil scientists see the microbial activity of soil as having a larger effect on plant health than the mineral make-up of the soil itself. This is logical as the speed of a given soil's ability to break down organic matter is based largely on its microbial content: the more matter broken down by micro-organisms, the more plant-accessible nutrients replenish the soil. Microbes are often overlooked, and in the context of terroir are rarely discussed.

Smelly microbes

The earthy smell in wine can be attributed to a molecule called geosmin, which is part of the family of terpenoid oils that are the main bioactive ingredient of essential oils. Geosmin has a pungent, 'earthy' odour, and many animals are hypersensitive to even the smallest quantities of it – humans can detect it at levels of just 100 parts in a trillion! It's created by a soil-dwelling bacteria called *Streptomyces*. Its strong smell attracts small soil animals which are then able to spread *Streptomyces* a little further afield. Geosmin is also present in beetroot, spinach and mushrooms, which is why they can often smell quite earthy. We will discuss how terpenes like geosmin get into wine a little later in this book.

Petrichor is the wonderful term we use for the most delightful of smells that follows, quite specifically, the first rain after a dry spell. During dry conditions, plants release stearic and palmitic acid in the form of yellow-looking oils which accumulate in between rocks and soil. The yellow oils are then broken down by soil bacteria. When it rains these broken-down residues and geosmin are made volatile and dispersed into the air to create the refreshing earthy smell of petrichor. Stearic and palmitic acids cause plant leaves to wilt and wither on contact and plants are thought to produce them as a means to weaken neighbouring plants during dry conditions and gain a competitive water advantage.

While we are on the subject of soil bacteria I just wanted to take a small tangent from wine into general health. Recently, there has been much debate about people's proximity to nature and links to reduced rates of depression and anxiety. In 2004 an oncologist by the name of Mary O'Brien (based at the Royal Marsden in Surrey, England) was investigating a relationship between a soil bacterium called *Mycobacterium vaccae* and the immune responses from her lung cancer patients. The soil bacterium had proven anti-inflammatory effects, so she ran a study using a high concentration serum. To her dismay, the subjects injected with her serum did not live any longer but what was notable was that they had consistently reported feeling a bit happier. Since her initial discovery there's been much extended research on the subject with neuroscientist Dr Lowry of Bristol University hypothesizing direct links to raised serotonin levels for those treated with the bacteria and proposing it as a treatment for the side effects of PTSD. Links between geosmin and other soil bacterium levels in wine and wider happiness are yet to be investigated, I'd be curious.

Humus

The soil science term 'humus' denotes the organic material in soil. Humus is mostly made up of decomposed plant and animal matter, which can include decayed manure, moss, lichens and vegetable matter. It is a core part of plant nutrition because of its phenomenally high CEC, even higher than that of any of the clays we mentioned, due to the tiny organic particles humus is made up of. Humus also plays a pivotal role in maintaining soil health, structure and texture, yet we never see mention of it anywhere on a bottle of wine.

Soil structure is the way different soil components are held together. A good soil structure would ensure soils are neither easily compressed nor too loosely held together; a healthy balance needs to be struck. Compressed or compacted soils have low porosity and result in poor water and air saturation. Poor water saturation restricts the exchange medium which enables cation exchange to occur, and if gases cannot permeate soils then we run the risk of suffocating the development of microbial communities. Both of these effects will mean a reduction in soil nutrient availability and fertility.

Vine roots need around 10–15 per cent of the soil to be air-filled space. If soils are too porous and the large cavities in the soil can easily fill with water this may lead to vine death. Waterlogging damages vines via the displacement of oxygen by water. Vine roots require oxygen to respire, and many tiny soil micro-organisms also require an element of oxygen to prosper, so without enough available oxygen both roots and microbial populations will suffer.

As vines grow, more nutrients are required to sustain a now larger and hungrier plant. During the growing period it is important that vines don't suffer water stress as it will negatively affect their development. The term 'water stress' refers both to a lack of water and to waterlogging. Water is the nutrient-carrying medium; too little cuts our nutrient supply but too much kills our nutrient-creating little critters.

The term 'texture' in relation to soil relates to the average size of soil particles. Sandy soils will feel more porous than clay soils, which will smear to the touch. Soil porosity is an important factor in establishing drainage properties. Thinking about the two soils above, sandy soils will drain quicker as they'll let water fall right through; waterlogging won't be an issue but they will retain a lower nutrient storage capacity as a result. Clay soils may cause waterlogging if there's an overabundance of the wrong sorts of clay, and high rainfall, though in drier seasons they may prove more fertile. It is a common theme in nature – balance is key and loam soils often prove better on all fronts than soils dominant in either clay or sand.

"Humus has far more of an effect on nutrient availability than geology, yet we never give it the limelight it deserves."

Humus acts as the glue that binds it all together, and it needs to be replenished regularly as micro-organisms in the soil are constantly working to break it down. In the process of microbial decay, gas bubbles are almost always a by-product, increasing porosity and the release of nutrients back into the soil. Many vineyards practise 'topping up' by adding manure or compost to soils or, in the case of conventional farming, synthetic fertilizers. We've focused a lot on CEC within the context of minerals, and it's easy to overlook the more obvious truth: an overwhelming quantity of nutrients are not supplied by rocks or minerals, but by decaying organic matter, humus. Humus has far more of an effect on nutrient availability than geology, and further direct benefits to soil structure and water drainage, yet we never give it the limelight it deserves. Humus never gets credited on a wine bottle or in wine writings.

Furthermore, our underappreciated humus is uniquely nitrogen rich. Earlier, we learned that nitrogen is important for plant growth: it's a key requirement for vegetative growth and the building block for protein development in plants. Rocks and minerals can only supply trace amounts of nitrogen, nowhere near the volume needed for plants to grow. Humus comprises about 60 per cent carbon, 6 per cent nitrogen and smaller amounts of phosphorus and sulphur. It's a far better source of nitrogen than rocks and the main reason we've been chucking poo on our crops for centuries.

In terms of how this affects our wine, in a growing environment with no external chemical addition, links have been made between the application of humus and the final colour of red wine. Vineyards with more humus will contain more nitrogen, increasing the green canopy growth of vines which, without human canopy management, will shelter the grapes from direct sunlight. This creates lighter purple-hued red wines as opposed to those exposed directly to sunlight, which present as darker reds. Of course, at every stage of production this can be combated by human intervention, from fertilizer addition to clever pruning in the vineyards, but it is nonetheless a fascinating chain of events.

Now we've discovered that the living soil organisms and humus have a greater effect on soil health than geology, shouldn't our conversations at the table be about them? Rotting dung and terroir-specific colonies of bacteria don't quite carry the romance which talk of geology and stories of place do, but they do undoubtably play a bigger role in viticulture.

Soil acidity

Let's transport ourselves back to ancient Egyptian times. Along the banks of the Nile, towards the mouth of the Mediterranean Sea, salt-loving marsh plants grow in abundance. City dwellers used to burn these saltwort plants for numerous purposes; the ashes were dubbed 'al-quili' and the wind would spread them far and wide. The Nile also took its fair share of al-quili, transporting it out towards the sea. Downstream, locals washed their wares and noticed some quite extraordinary effects. Their technicoloured garments would change colour entirely: bright reds turned to greens, greens to blues, and blues to purples. The Nile was deemed responsible for many a miracle, and it took a couple of thousands of years before we managed to unlock the science behind this seemingly magical transformation.

An acid refers to any substance that contains hydrogen and is capable of donating a proton (H^+) to another substance. In alkalis, the opposite occurs and hydroxide ions (OH^-) are produced instead; these neutralize acids. The word alkali derives from the Arabic word 'al-quili' – we now know that these burned ashes formed a powerful alkali that tinkered with the natural dye used in clothes when they were washed in the river. It has similar effects on our wine too.

In 1909 the Carlsberg Brewery in Denmark was researching how hydrogen ions affected enzymes in beer. Due to the sheer numbers of ions they were dealing with in the realm of acidity, their measurements needed to be performed on a logarithmic spectrum, and thus the pH scale was derived. In a neutral liquid such as distilled water, neither kind of ion is available for reaction, and it will have a pH of 7. Anything below this on the scale will be an acid, and anything above an alkali. Because of the logarithmic aspect to the scale, a weak alkali with a pH of 8 is ten times weaker than an alkali with a pH of 9, and one hundred times less alkaline than one with a pH of 10.

It is often posited in wine that the parent bedrock of soil influences its pH. But this cannot be the case and soil pH is more likely to be a factor of climate, land use or living soil organisms. The one major exception to this statement is calcareous rocks. Calcareous rocks include limestones and contain calcium carbonate. The carbonate binds with H^+ ions, which in turn reduces the concentration of free H^+ ions in the soil, increasing the overall pH and making the soil less acidic.

How acidity affects vines

Specific grape species, grape varieties and rootstocks react differently to soil acidity. Some will thrive in lower acidity soils, where others may struggle. The acidity of a soil has a pronounced impact on microbes and other soil life, as well as ion solubility.

Microbial and underground biological activity. Microbes and soil critters are sensitive to their environmental acidity. Soils with pH levels between 5.5 and 7 tend to confer better health and vitality upon the soil life within, which in turn directly links to soil nutrient availability, texture and porosity.

Ion solubility. Different nutrient cations will dissolve in water better at differing degrees of acidity. You can see why an understanding of soil science is a full-time job! Examples include:

- Nitrogen is most available in water between pH 5.5 and 8.5 and tails off outside this range.
- Manganese is most water soluble between pH 5 and 6.5.
- Boron has two sweet spots, the first between pH 6 and 7, peaking again at pH 8.7–10.

Many nutrient deficiencies in vineyards can be rectified by taking measures to correct soil acidity. To get the best average nutrient solubility across the board for most nutrient cations we're looking at around the pH 6.5 mark, a mildly acidic soil.

Soil layers

Soil can be divided up into four main layers. From the surface down these are: topsoil, subsoil, weathered rock and bedrock. Nutrient ions are largely found in the topsoil due to the weathering processes we discussed earlier, and since topsoil is likely to be higher in oxygen and humus content than lower layers it's teeming with micro-organisms and nutrient-providing lifeforms. The deeper down into the soil we go, the less populated it is with organisms and nutrients.

This top layer of soil is also the layer that has been transported the most. Many topsoils have travelled far and wide through the action of wind and water erosion and as a result have little mineral make-up in common with the slow moving bedrock directly below.

The weight of each soil layer above means that the deeper we delve, the more compacted and dense the soil becomes. This also results in lower microbial activity the deeper we go. In extreme cases, compaction can lead to soils being so tightly packed at deeper levels that water cannot drain. Water may then build up above the tightly packed lower layers, creating a slush of water and soil; this instability effect is referred to as 'liquefaction'. In sandy soils we more commonly dub this 'quicksand'. These effects can be caused by the overdevelopment of land and movement of heavy machinery above ground.

Underground rivers and water aquifers are water-rich parts of the soil. The relatively clear, flowing rivers we find above ground are entirely different from the muddy ones we find beneath the soil. Groundwater ebbs through underground rivers, carrying soil particles with it in a rich, earthy sludge. The average sitting time of groundwater in an aquifer is over a century and it will contain many dissolved minerals within. When we drink mineral water, we consume all these nutrients directly. We know vine roots are selective with these nutrients, and even if taken up by the vine, not many will end up inside a grape. Even if found in the grape, many will be destroyed or converted after fermentation and wine production. Following this line of logic it becomes evident that little of what makes up a soil is going to be directly translated to our glass of wine.

When talking 'terroir' with science-minded grape growers (often following regenerative or organic principles), it's the porosity of the soil which I'm often told makes a dramatic difference. Remarkably, it was an ancient oak tree in London's Kew Gardens that taught the world a lesson in soil porosity. The year 1987 was a fateful one, not only because it was the year of my own birth, but also because it saw one of the biggest storms the south-east of England has ever seen, aptly named 'the great storm'. The winds ripped down more than 15 million trees in the space of an hour, among them the 200-year-old Turner's Oak in Kew Gardens. This huge oak tree had been quite sickly for some time and many assumed it was nearing an end well before the storm came in and lifted it up metres above the ground. The tree's roots were exposed to the air for mere moments, before it was thrown violently back to earth. Soon after the storm, tree surgeons began the arduous task of making safe and cutting down the hundreds of other smaller trees in the Gardens which had fallen. The great Turner's Oak was last on the chopping block; it took three years of chopping to eventually arrive at the great oak. When its sentence became due Tony Kirkham, head of the Arboretum at Kew Gardens, witnessed something that completely changed our understanding of soil. The Turner's Oak was a picture of health, and in better shape than it had been even before the storm hit. Tony realized that over decades of crowds and visitors to the gardens the soils had compacted around the oak, pushing out oxygen and water and slowly suffocating its microbial population. The storm served to introduce fresh oxygen and porosity to the tree's root plate, breathing new life into the soil surrounding the tree, and led to a 30 per cent increase in new growth over the few years that followed. He also realized that the same issue had affected every other tree in Kew Gardens, leading to changes in the way Kew Gardens manages its trees. In fact, this discovery catalysed a change in tree management the world over. Nowadays trees in high population areas may have their soils revitalized by air spades that inject the earth with freshly pumped air, increasing porosity and keeping their microbial communities alive.

Terroir

Alice Vacani's Oxford University thesis, 'The Evolution of Oenological Chemistry in Winemaking', reports that the term terroir originated in medieval Burgundy. It became commonplace for repenting bourgeoisie to give gifts of land to the monastic community on their deathbeds in a desperate attempt to seek atonement. The monastic order soon began to take advantage of this practice and, realizing that some parcels of land led to tastier wines than others, began to request specific plots within estates. This led on to the *cru* system and our modern definition of terroir.

It is currently agreed that terroir is a combination of climate, soil, terrain and tradition (winemaking practice). Soils are complex environments to navigate; aside from learning the core principles of soil science, we need to bear in mind that the make-up of our soils varies hugely from one square metre to the next, and so do the wines produced from them.

Terroir is often described as a 'sense of place', and it's saddening to realize that the two components that make the largest impact to plant life in soils – micro-organisms and humus – aren't ever included in this sense of the term. Microbial prejudice has run rife for centuries. Some late twentieth century vintages of Beaucastel Châteauneuf-du-Pape were highly praised by famed tasters and wine writers of the time. They commended the wines and the famed terroir they hailed from but years later it was found that their unique terroir-specific character turned out to be a strong case of *Brettanomyces* yeasts which had transformed the wine.

Taking a more cynical approach we can look to how wine is marketed. In practice we would struggle to sell a wine off the back of mycorrhizal soil bacteria. It is far more effective to spin tales on aspects of the local landscape that have been with us for millennia.

Let's look now for correlations that can be found linking geology and wine style. Limestones vary massively but they are all calcareous and do consistently provide well-drained and high pH soils; in wine literature they are said to produce wines of freshness, minerality and bright natural acidity. But where's the link from what we've learned to concepts of freshness, minerality and acidity in the glass? It's also worth noting that much wine literature also proposes the exact opposite, claiming fuller bodied burgundies often originate from limestone soils. Can these opposing statements both be true? We in wine should be a bit more careful when using such assertions in tastings or on the back of a wine bottle.

One of the most influential studies made about the impact of soil on terroir was conducted by Professor Gerard Seguin in the mid-1990s. He was a French scientist at the University of Bordeaux who concluded that no reliable link could be formed between the material composition of the soil and wine characteristics or quality. He stated: 'Excellent wines can also be produced on acidic, basic and neutral soils. Excellent wines can also be produced on soils with balanced chemistry, and on those with nutrient deficiencies.' He found that it was the availability of water during crucial vine growth periods, as well as the soil's texture and drainage properties, that really made a difference.

3.
Natural law

Natural law

We begin with a story, one that all with an interest in food and drink should be familiar with, since it transformed the world we live in. This story ties in perfectly with the principles of soil science we have learned so far. Having delivered numerous talks on the subject myself, I feel the more we discuss it the better our chances of making a positive change to the food and wine we bring to the table.

Fritz Haber

The origin story of synthetic fertilizer is a tale of Oppenheimer-scale controversy and sadness, one which paves the way for how we currently farm. Before the invention of synthetic fertilizers by the German-born Jewish scientist Fritz Haber, famine was a real and worldwide issue. With an ever-growing global population, the crops we farmed were not producing the yields required to sustain our increasing numbers. The issue, as we now quite clearly understand, was the need for more nitrogen. We learned from the first two chapters how nitrogen is a key growth nutrient and that has a direct impact on yields. We also discovered that even though there's an abundance of nitrogen in the environment, bacterial cooperation is required to release just a small amount in a form that plant life can absorb. Since the emergence of farming, whether early civilizations realized it or not, we've always been on the lookout for additional sources of nitrogen. Guano (the poo of seabirds and bats) provides a concentrated source of nitrogen, potassium and phosphorus. Seabirds and bats have a much higher protein-based diet of insects and sea life than other common land animals, hence their dung is a far more effective nitrogen-rich fertilizer. In the 1850s the cost of guano hit an all-time high at a quarter of the cost of gold – wars were sparked over it. The Chincha Island War in 1865 between Spain and the guano-rich islands of its former colonies of Peru, Ecuador, Chile and Bolivia were all fought over bird poo. There was such a demand for guano that the USA devised the Guano Island Act in 1856. This enabled US citizens to take possession of unclaimed and unoccupied guano-rich islands in the name of the United States, and once claimed, permitted the government to use military forces to protect them. The struggle was real, and people were desperate.

Alongside fellow chemist Carl Bosch, Haber created the Haber–Bosch Process in 1909. They devised a cunning way to extract nitrogen from air, in which it is the most abundant element, and convert it to ammonia, a soluble form of nitrogen which plant life can take up. This one process completely solved the nitrogen issue, opening the door for the world to take the first steps out of global famine and into a world of abundance.

A couple of years later, in 1911, Haber was appointed as the Director of the Institute for Physical and Electrochemistry in Berlin, and in 1914 the First World

War began. Haber was a patriot, and he soon turned his research from agriculture to weapons of mass destruction in order to best support his country. He began testing pesticides as chemical weapons. On 22 April 1915 the first large-scale use of lethal poison gas took place on a battlefield in Ypres, Belgium, supervized by Haber. A chlorine-based gas was pumped over enemy trenches; the heavy gas sank into the trenches and suffocated the opposition forces. A month later Haber's first wife Clara, a fellow chemist, took his military pistol at night and shot herself in the chest: she passed away in the arms of their only son. The following morning, Haber led the first gas attack against the Russians on the Eastern Front. There's much speculation on what had led his wife to suicide but the morality of her husband's research in both chemistry and war is often seen as the key driver. Haber further threw himself into his work, and just a year later was appointed Chief of Germany's Chemical Warfare Service.

The 1918 Nobel Peace Prize in Chemistry was initially awarded to Haber for 'the synthesis of ammonia from its elements'. Given his ongoing work in chemical warfare there was much controversy around this decision and the prize committee decided that none of the year's nominations met the criteria as outlined by Alfred Nobel in his will. They withheld the award in 1918 and ported it to 1919, when Haber inevitably received the prize.

In 1933 Hitler came to power. Anti-Jewish decrees made Haber's position untenable, and he had no choice but to flee the homeland he had devoted his life to. He passed away a year later due to heart complications.

Sadly, the story doesn't quite finish with Haber's passing. One of Haber's early pesticides and rodent killers based on cyanide was developed in the First World War for use as a poisonous gas in warfare; called Zyklon, it was banned shortly after the end of the war. In the 1940s Nazi scientists tinkered with Zyklon, taking the chlorine smell component out of the chemical to create Zyklon-B. This gas relied on cyanide to interfere with human cellular respiration, preventing cells from producing ATP, a chemical essential to the functioning of many biological processes. It went on to be the main gas used in the gas chambers of concentration camps, spelling the genocide of over 1.1 million Jews by use of this gas alone. An outcome Haber could never have foreseen when first looking to science for a cure to world hunger.

How we farm now

The Haber–Bosch Process completely altered the landscape of farming on a global scale. Each year 500 million tons of artificial fertilizer are produced using this process, and the crops grown sustain 40 per cent of the global population. Humans first farmed more than 12,000 years ago, and archaeological evidence has shown that farming methods changed very little prior to this twentieth-century discovery. In the last 70 years the way we farm and produce food has changed dramatically from how our ancestors did it.

The twentieth century's two world wars brought about much global change, and the monumental alterations in food and drink production are little dwelled upon. Today's farming relies heavily on artificial fertilizers, fungicides and pesticides to achieve large and reliable yields. Food costs are lower than ever, spurred by the emergence of supermarkets which have allowed companies to sell in large quantities around the world. Supermarkets operate as one would expect, keeping costs low and driving demand. This puts pressures on farmers, which in turn creates higher requirements for use of synthetic chemicals in order to keep food costs and wastage down.

In the last three decades much evidence has surfaced illustrating the harm caused by what we now call 'conventional farming'. Between 1993 and 2009 the European Commission (EC) carried out major reviews on pesticides. Over a thousand pesticides were investigated, and more than two thirds of them were consequently excluded from the market. This movement by the EC was called the 'big clean-up', and the 250-odd pesticides that were approved by the EU were included because the investigation concluded they had no harmful effects on consumers, farmers and passers-by and they did not cause 'inordinate' damage to the environment. We still battle with the use of agrochemicals in farming, new chemicals are being devised, and the banning of newly developed harmful ones is an arduous process despite our knowledge of the negative effects which surround them. More than 80 new substances have been added to the list of approved pesticides since 1989.

What are agricultural fungicides and what impact do they have?

On 23 November 2023 this question was posed in the House of Lords (UK Parliament). Within the debate that followed, research on synthetic pesticides, fertilizers and fungicides from the Parliamentary Office of Science and Technology (POST) was brought to the table, highlighting why we need them:

'Pesticides help ensure food availability and affordability by enhancing crop productivity, improving appearance of produce and maintaining food safety. By preventing pest damage, moulds and toxins, they extend the shelf life of food and reduce food spoilage and waste.'

Within this same discussion the amount of pesticide use necessary to provide food for a growing global population was also called into question. Some academics, farming groups and charities have suggested that reduced or minimal use of pesticides (such as in organic agriculture), combined with changes in diet and reduction of food waste, could produce enough food for the population. Opposing academics stated that this would require more land and may potentially increase greenhouse gas emissions. They, together with other farming stakeholders, maintained that a reduction in pesticides would reduce food production, affecting food availability and affordability.

This debate also highlighted some of the environmental and health issues that have arisen from the use of agrochemicals:

- Fungicides can affect the microbial gut fauna of invertebrates, such as honeybees, although little is known about their impact on insects other than pollinators.
- Pesticide pollution drives biodiversity loss in Europe. It causes significant declines in insect populations, threatening the critical role they play in food production.
- The widespread use of chemical control products such as fungicides can drive problematic traits such as fungicide resistance.
- Agricultural fungicide runoff can affect freshwater environments when entering watercourses. It added that pesticides can be washed into water bodies by rainwater or may enter them directly if sprayed close to water. It said pesticides can also enter groundwater via soil infiltration.
- People can be exposed to pesticides indirectly through their environment. Farmers are not required by law to notify people nearby when spraying is taking place, although this is regarded as best practice and voluntary initiatives encourage this.
- A large-scale human biomonitoring study conducted between 2014 and 2021 across five European countries found that at least two pesticides were present in the bodies of 84% of survey participants. Pesticide levels were consistently higher in children than in adults.
- According to the European Environment Agency, strong or suspected links have been established between human exposure to chemical pesticides and an increased risk of chronic illnesses such as cancer and heart, respiratory and neurological diseases.

"Environmental and health issues have arisen from the use of agrochemicals."

The harm to human and environmental health is clear and reported from multiple sources across the world. Controversial chemicals like glyphosate, dicamba and chlorpyrifos are still used routinely even though a litany of life-changing human health issues have arisen around their use. These chemicals are produced by billion-dollar businesses like Monsanto and Bayer who have far-reaching influence and will do all they can to keep the cogs of their businesses turning.

The further damage that chemical fungicides pose to soil life and mycelial networks isn't often discussed and it is certainly not brought up in matters of Parliament. The damage is measurable, and without mycelium or living microbial soil populations like the *Azotobacter*, we know that our soils will simply stop regenerating. The use of these synthetic agrochemicals effectively kills soil life, rendering soils unable to generate nutrients and thus heightening their need for further chemical application.

In viticulture specifically there have been some *Vitis vinifera* related hindrances which have further increased our dependency on pesticides and fungicides. Around the time we overcame phylloxera in the 1800s, two other notable diseases arrived in Europe. Powdery mildew is a fungal disease that was first spotted in French vineyards in 1847, just two years after phylloxera took hold. It's a plant pathogen that also originated in North America and forms a powdery-looking fungal coat over plant leaves. Powdery mildew is best combated by use of sulphur in the vineyards. Downy mildew is another fungal infection transported from North America; it was first reported in vineyards around 1878. We use copper to combat downy mildew, most commonly in a concoction called Bordeaux mixture. These diseases are responsible for the wine world's biggest farming dilemma. Like conventional farming, viticulture has come to rely on these applications and practices to overcome mildews and ensure its financial security in the global economy.

"The use of these agrochemicals effectively kills soil life."

Common practices and applications used today

Bordeaux mixture

This preventative formula was developed in the late nineteenth century, in Bordeaux, which has a climate very conducive to mildews. This effective blue fungicide is made up of copper sulphate, quicklime and water. The active copper ingredient denatures the enzymes in fungal spores, preventing them from spreading and germinating. This preparation has now been banned in most parts of the EU since copper ions can build up in soils leading to heavy metal pollution and toxicity, and can also accumulate inside living organisms like earthworms, drastically decreasing their poulations. It also alters the acidity of soils due to the alkaline properties of quicklime. After a bout of rain, think of the havoc this is likely to wreak on those intricate and all-important mycorrhizal networks below ground.

Bordeaux mixture is banned in 18 of the 27 EU countries; it is still permitted in Italy, Portugal, Greece, Cyprus, Bulgaria, Romania, Slovenia, Hungary and Malta. Outside of the EU, new wine regions like the UK permit its use in commercial farming, and large wine-producing regions like Australia continue to spray it.

Both copper sulphate and quicklime are naturally occurring ingredients, hence can be marketed as 'organic' and used on certified organic farms!

Copper

Aside from Bordeaux mixture, other copper compounds are often used as fungicides in farming, since it is a powerful antimicrobial agent. There's much controversy about its use in organic farming – it is permitted since it is viewed as a non-synthetic chemical but it clearly presents dangers to all fungi, whether friend or foe, which would seem to set it against the principles of organic farming. Sadly, there's no current commercially available or effective alternative to copper. It is still used extensively in agriculture, and all that can be done is to minimize its use rather than get rid of it altogether.

NPK fertilizer

If the Haber–Bosch Process is the goose that laid the golden egg, NPK is said egg. This blend of nitrogen (N), phosphorus (P) and potassium (K) plugs many of the gaps which nature fails to fill. Nitrogen boosts green growth in plants and is most needed in spring and early summer, when plants require this push the most. Phosphorus promotes root development and is applied on seed sowing, and potassium helps flowering and fruit-set so is most needed towards the end of spring and into early summer. NPK has become essential to nutrient supply in conventional mass farming. We can now produce tremendous yields, but when paired with the use of fungicide and pesticide, our soils lose their ability to regenerate. This then leads to the need for increased NPK application the next year and the year after, and so on. It is an addictive cycle and we're very much in the eye of the storm with little viable solution to maintaining yields at current levels.

Nitrogen leakage into watercourses promotes the eutrophication of lakes, encouraging the overgrowth of aquatic plant life at the expense of the animal life of the lake. This well-documented problem arises mostly from the use of synthetic fertilizers, which are more soluble and fast-releasing than their natural counterparts.

Tilling

Tilling and ploughing have been commonplace practices in farming for thousands of years. We've done this to aid the mixing of organic humus, which settles on the top layer of soil, into the next deeper layer, it helps us in planting seeds, and shreds up weed networks to assist our planted crop in gaining a competitive advantage in the early stages of its growth. We've always had good reason to do this, but of late there's been much controversy around this practice. 'No-till' is a movement which brings to light some of the new scientific discoveries which illustrate the damage it does to soils. Tilling the land has been proven to both disrupt soil structure and destroy mycelial exchange networks. In addition to this it accelerates surface water runoff and soil erosion, resulting in a net reduction in soil fertility.

In the 1930s we learned the dangers of continuous land tilling the hard way. Swathes of North American and Canadian prairies were hit by the Dust Bowl incident, a period of severe dust storms which, at their most extreme, blocked out the sun. This wreaked ecological havoc and made over half a million Americans homeless in the process. The phenomenon was caused by a natural drought, paired with the systematic practice of deep ploughing the topsoil of the Great Plains over the preceding decade. The tilling caused unanchored soil to dry, turn to dust and get whipped up into the air. This incident was a wake-up call for environmental farming practices and, with the backing of President Roosevelt, Hugh Hammond Bennett 'the father of soil conservation' led a campaign to reform farming practices. He introduced methods of farming that we use today such as crop rotation, strip farming, planting cover crops and leaving fallow fields. Yet there was still local resistance from farmers despite the visible chaos that tilling caused. The government went a step further by subsidizing farmers by a dollar an acre in the places new farming methods were adopted. In addition, Roosevelt created the Civilian Conservation Corps, which ordered the planting of 200 million trees from Canada to Texas, serving as a windbreak to help hold the soil in place. Despite this dramatic lesson, tilling is still common practice around the world.

It all sounds quite bleak. How are we to sustain the numbers of people on this planet, and our buying habits, whilst eschewing the conventional farming practices that have been proved to destroy our soils? As our population has grown, most of us have separated ourselves from the spaces which cultivate our food, and grown distant from our knowledge of what goes on in food and farming. On a more positive note, perhaps we are at a crossroads where, if we take the right path we may be able to remove our reliance on conventional farming. So what other options are there?

Organics

It's become *the* buzzword in marketing food, drink and even clothing across the developed world. Many wine fanatics and professionals remain sceptical about the quality or even the idea of organic wines, yet few of us know exactly what organic food or wine really means. We may know what the word means, and make some assumptions on what that means for wine, but only by knowing what the term means legally can we begin to comprehend how it can or can't affect our wine, and if we deem it important.

The EU has defined organic growing thus: 'An agricultural method that aims to produce food using natural substances and processes ... organic farming tends to have a limited environmental impact as it encourages responsible use of energy and natural resources, maintenance of biodiversity, preservation of regional ecological balances, enhancement of soil fertility and maintenance of water quality.'

The European Union regulations on organic farming are designed to provide a clear structure for the production of organic goods. Certification was created to satisfy a consumer demand for trustworthy organic products whilst providing a fair marketplace for producers, distributors and marketers. Confusingly within organics there are many different certifications which tend broadly to cover geographic remits, and all have to yield to the International Federation of Organic Agriculture Movements (IFOAM). Many organic growers find IFOAM far too lax, so adopt more stringent measures at their wineries. How are we as consumers to separate the organizations that practise organics to basic IFOAM requirements from those growers who adopt a more stringent set of rules?

Types of organic bodies include Nature et Progrès (France), Agriculture Biologique (France), Jas (Japan), Ecovin (Germany), the Soil Association (UK) and many more. These examples of organic certification bodies range in stringency, some stricter than others, yet all need to meet the basic standards of IFOAM.

All organic vineyards must go through a period of conversion, a stepping into organic standards which takes a minimum of two years. Many winemakers comment on how this is not long enough, saying that getting rid of previously sprayed residues in the soil can take more than ten years. Certification has a price tag for the grower so adds a cost to the wine, as do the conversion period and the reported slightly reduced yields. On the plus side there is the added cachet among certain sectors in having your wine labelled organic. Europe is an organic market leader, home to 84 per cent of the world's organic viticulture. In 2021 it was reported that close to 8 per cent of vineyards worldwide are classed as organic, more than five times that recorded in 2005. There's a fair bit packed into what it means to be organic, and we'll split it into two parts: the rules that apply to farming, and those on production.

"How are we to sustain the numbers of people on this planet whilst eschewing the conventional farming practices that destroy our soils?"

Organic rules on farming (IFOAM)

From the first day of the organic conversion period the vineyard must run as a fully organic vineyard would. This means prohibiting:

- Synthetic fertilizers (like NPK)
- Synthetic chemical pesticides, fungicides and herbicides
- Genetically modified produce

The following are permitted:

- Organic fertilizers derived from animals (manure, guano), plants (humus, algae), or minerals formed from animal or plant matter (limestone, natural potassium sulphate, calcium sulphate, etc)
- Copper sprays
- Sulphur sprays
- Use of other plants, animals or micro-organisms in the vineyard.

There's an entire list of organic fertilizers that IFOAM allows farms to use. These are broadly less soluble than the synthetic stuff so the risk of run-off into unwanted places is reduced. With certification comes bureaucracy. It's impossible to avoid, and by definition a certification will need to have an element of box ticking to it. Everything needs to be checked, otherwise the merit of the certification degrades. All organic plant protection and fertilizers need to be on a pre-approved list or go through an approval process. The list already includes the commonly used applications, giving specifications for manure, liquid animal excrements, guano and so forth; however, many other preparations or natural remedies may not be permitted in organics without a painstakingly long and pricey approval process. The 'nettle manure war' in 2006 was sparked by an instance where French grower Eric Petiot had his computer impounded and was heavily fined for publishing a book on how to make your own nettle water fertilizer and pesticide for organic farming. Nettles are found growing just about everywhere, and his recipe was quite easy to knock up at home. However, the nettle concoction was not approved at the time of publication, and legal repercussions followed. Years later the EU agreed a recipe if it is sold through approved channels to ensure it is made correctly. It is easy to look cynically at this case and think the EU is monetizing a readily available product, but

No GM

Organics has a blanket ban on genetically modified (GM) products; from vine to yeasts, none of it can be GM. The argument behind banning GM is purely a philosophical one: they are not strictly from nature so cannot be organic. Keeping in mind that the EU's definition of organic is wholly about limiting the harm to the environment, permitting the use of GM could realistically assist in achieving these principles. GM crops are not made with the intention to damage the environment, and are often designed to be tougher, stronger or faster than their natural counterparts. Further research into GM vines could help us develop powdery or downy mildew resistant varieties which may not require any copper and would easily reduce the requirements for synthetic agrochemicals (like many of the hybrid varieties can do). The issue with GM plants is that they are a relatively new phenomenon and we do not have enough information on how they may affect the environment and our bodies. It is perhaps understandable that, at least for now, organic farmers would want to keep them out of the certification.

should we, the consumers, trust any old homemade concoctions farmers make up? Let's not forget the whole point of certification is to port the consumer's trust of the individual farmer to that of a governing body, and if we are to trust the governing body we can't have it taking any risks. It is not an easy situation to resolve: as a result of the distance between farmers and consumers, consumers do not see how their food is farmed.

IFOAM also mentions the use of other plants, animals and micro-organisms. In essence we're talking biodiversity, with the aim of encouraging symbiosis in the vineyard, and there are many ways to do it.

So, we're still allowed to use copper?

There is much controversy surrounding the use of copper in organics – it's a heavy metal, toxic to soil and water in higher concentrations, yet it's not synthetic and can be used in organics. Copper is a microbial surface killer and destroys microorganisms in many ways. On contact it degrades the cell walls and membranes of microbes via the flow of free radicals and electrons that naturally occurs when copper reacts with oxygen. This weakens microbes and renders their cells susceptible to death by copper ion. Since copper is also toxic, whilst oxidizing, toxic copper ions flood into microbe cells and destroy all the fragile and important innards like their DNA-making parts. Not a pleasant outcome for our microbial friends or foes.

Organic farmers are more dependent on copper than their conventional farming counterparts, compensating for the synthetic chemicals they are not permitted to use. Growers comment that copper is even more damaging to life and responsible for more water contamination beneath the soil than many of the synthetic chemicals. Until there is a reasonable substitute to copper, there will be little change apart from a careful 'reduction where possible' approach. For organic farming in the EU limits on copper have been put in place. In 2003 a maximum of 8 kilograms per hectare were permitted annually; this was further reduced in 2006 to 6 kilograms per hectare, or more precisely 30 kilograms per hectare per five-year period.

More stringent restrictions have been looked into by the French government agency Agence Nationale de Sécurité Sanitaire de l'Alimentation, which aimed to further reduce this to 4 kilograms annually but ongoing studies deemed the reductions too damaging to yields in years of heavy downy mildew attacks. Hence it remains steadfast at 6 kilograms per hectare.

There are also additional pressures for organic restrictions to become more relaxed. Organic governance must give all applicants a fair chance of farming organically, from wet and rainy regions, to the warm and dry. For it to be all-inclusive, the rules need to be relatively generous. Germany is full of rivers, has high humidity and suffers from inconsistent sunlight. It's a tough organic climate to work in because of the elevated risk of fungal threat and it's no surprise that German wine growers are disputing the restrictions on copper altogether. If Germany is to have an opportunity to produce organic wines at all, they have to push back. This, then, would widen the remit for all organic rules across the EU. Push back encourages more countries and producers to sign up to organics, yet waters down what many consumers may expect their organic wine to mean.

"Copper is damaging to microbial life and responsible for more water contamination beneath the soil than many of the synthetic chemicals."

Nitrogen fixers

This is a clever way of utilizing plants which help absorb nitrogen from the air and sequester it in the soils in a form other plants can take up. Leguminous plants are renowned for this sort of work: lucerne and clover have commonly been adopted for the purpose in farms around the world. However, since the development of the Haber–Bosch Process they've been sidelined. Do note we've now discovered that it's microbes that do the bulk of the nitrogen-fixing work, and extensive use of copper will inhibit these little microbes from working at their best.

Cover crops and weeds

Organic growers aren't allowed to use chemical weed control and can only take them away by mechanical means. Planting shallow rooted cover crops like grass or clover is a brilliant way to create competition for vineyard weeds and in turn reduces the risk of heavier resource sucking weeds. Choosing the right cover crops is crucial, and finding suitable partners will have impressive effects on reducing soil erosion and improving structure, assisting in water retention and promoting microbial soil life, all of which improves soil nutrient levels. In high water and nutrient locations, greater competition may be good for the vines, creating the conditions to make life harder for them and pushing them into survival mode.

In addition to the plethora of positive points above, when the vigneron cuts the grass or clover it creates a brilliant supply of humus. Many studies suggest that planting cover crops also increases the development of mycorrhizal networks, which in turn will assist vines in making available more water, phosphorus, zinc, copper and iron.

Hedges

The simple hedge has been long overlooked and a good organic grower can talk your ear off on the subject. Many hedges were lost as technologies developed and mechanized agriculture took a hold in the 1960s. A simple hedge is teeming with life and provides food and shelter for a wide array of birds and other animals. They're a move away from monoculture and a step towards a biodiverse way of farming in a functional fashion. The bats, birds and little animals that live in hedges often eat the insects we're desperate to keep away from our vines, which in turn reduces the need for insecticides. Plus, like cover crops, they add to the humus content of the vineyard when they shed their leaves each year.

Sexual confusion

Aren't we all? Well, in this case it plays to the organic grower's advantage. Caterpillars often prove to be an issue in vineyards, causing havoc by munching away at valuable photosynthesizing leaves. Male moths and butterflies use pheromone scents to locate females, so by placing little artificial pheromone capsules around the vineyard, we can misdirect them. Less mating means fewer caterpillars, and safer vineyards for all without the use of pesticides.

Organic rules at the winery

The current designation of organic certification in the EU is a relatively new one where, from the 2012 vintage onwards, organic producers can call their wines 'organic' instead of just 'wine made from organically grown grapes'. A subtle difference, yet before this there were no regulations on wine production, additives or cellar practice beyond those put on conventionally farmed wines. Post 2012, all organically certified wines have the following limitations on winemaking and cellar practices which now need to be adhered to:

- Tighter sulphur additive restrictions for red, white, rosé and sweet wine styles. Across the board we're looking at a 7.5–33 per cent reduction in sulphur compared to their conventionally produced counterparts.
- No flash détente (heating the grapes at 95°C for several minutes with vapour and then submitting them to a strong vacuum to release flavour and texture compounds).
- Banning of lysozymes (enzymes used to bacterially stabilize high pH must).
- No vacuum distillation (forcing the wine through a distillation column at reduced pressure to extract specific compounds).
- No elimination of sulphur dioxide by physical processes.
- No electrodialysis treatment or cation exchangers to ensure tartaric stabilization of the wine.
- No partial dealcoholization of wine.
- No nano- or ultrafiltration.
- Banning of caramel addition as a colouring agent – this additive has been associated with heightened cancer risk.

The following winemaking practices are still permitted in organics:

- Heat treatments below 70°C.
- Use of ion exchange resins for rectification of concentrated must and reverse osmosis.
- Concentrating the must to increase overall sugar concentration.
- Addition of sucrose.
- Addition of dry and commercially produced yeast.
- The addition of fining agents such as isinglass, egg whites, tannins, edible gelatine and plant proteins from wheat or peas.
- The addition of tannins to increase the tannin content of wine.
- The addition of acacia gum (gum arabic) for tartaric and colour stabilization.

To add an element of complexity to the above, there are also more wine additions and practices that aren't permitted, dependent on the specific rules of the individual organic governing body. The severity of these restrictions depends on the governing body and how much stricter its own rules are than those of IFOAM. Other rules added by some bodies include restricting the use of enzymes, potassium bisulphate, aluminium bisulphate, ascorbic acid (vitamin C), using non-grape sugars, reverse osmosis, cryoextraction, even restricting the addition of lactic acid bacteria, acidification and many more!

It's not hard to see why people might feel confused or even slightly misled by organics: there's a lot to take in. With numerous intricate differences between certifying bodies, consumers would have to study the subject to fully understand the nuances behind what each organic certification does and doesn't permit. With many rules appearing arbitrary and some bodies banning methods that others permit it is hard to get a firm grip on what 'organic' technically means. Why is flash pasteurization forbidden, whilst reverse osmosis is not? Why are we allowed to chaptalize (add sugar to the must), when we can't use vacuum distillation to lower sugar levels?

What does organic mean in the New World?

Outside of the EU organic rules do vary quite a lot. The criteria are not consistent across all states of the USA, and some countries have no rules at all. Many US states have similar standards to those of the EU from a farming perspective, but are far tighter on the rules of vinification, some going as far as prohibiting the addition of sulphur entirely. The USA is one of the few countries where organic wines, as opposed to just organically grown grapes, have been around for a significant amount of time. It's a lot stricter, hence far fewer producers are able to meet their requirements.

One can imagine these stringent USDA (United States Department of Agriculture) organic rules might entirely change the style of wines produced, and we'll soon learn why wines labelled organic in the States may taste completely different from those labelled organic in the EU. It also means we are speaking a different language when we talk about organic wines in the US, compared to Europe. This may explain the uneasiness behind the talk of quality within the organic wine sphere, especially since removing sulphur entirely from production dramatically increases the risk of spoilage or changing the wine style in a dramatic way. This apprehension may well have made its way across the Atlantic, like so many cultural phenomena have done in the past, and may explain the fear of or distaste for organic wine from old-school wine drinkers.

Sustainable farming

For the many who deem organic viticulture to be a non-viable long-term option, sustainable farming may be the way forward. For quite some time the word 'sustainable' has become a catch-all phrase to mean everything and nothing; it's the new organic, and very in vogue. On wine labels the term 'sustainable' is not itself protected, however there are governing bodies much like in organics from whom sustainable farmers may get accreditation.

The UN defines sustainable development in the Brundtland Report of 1987 as 'development that meets the needs of the present without compromising the ability of the future generations to meet their own needs'. Sustainable farmers aim to keep the land and their surroundings as healthy as possible, without jeopardizing the profitability of the business. The French refer to sustainable viticulture, quite aptly, by the term *lutte raisonnée,* which translates to 'the reasoned struggle', a heart-warmingly existential look at sustainability.

Sustainable farmers can use synthetic products on the land when needed, whilst maintaining biodiversity, soil health, and may also take a view on other environmental impacts the business will have, such as power consumption, transport, water resources, waste management and employee wellbeing.

The more common sustainable accreditations seen on labels are Agriculture Raisonnée (France), Terra Vitis (France), TYFLO (Alsace), Integrated Production of Wine (South Africa), Sustainable Winegrowing NZ (New Zealand) and The Sustainable Winegrowing Program (California). Most of these certifications have no direct prohibitions on particular products and rely on limiting synthetic inputs to reduce the environmental impact. The main tenets in place for these sustainable certifications are recommendations to follow rather than rules (with the one exception of TYFLO in Alsace, which enforces a restriction on the amount of copper used).

Because of the broadness of the ideal, it's near impossible to know what a vineyard is doing if they call themselves sustainable and don't have certification. Without the term 'sustainable' being protected, anyone can pop it on a bottle willy nilly, thus certification is vital. Even with certification, many sustainable bodies have come under scrutiny for being too lax. New Zealand's Sustainable Winegrowing certification for example has had an abnormally high uptake for the scheme. Almost all commercial grape growers in the country have reduced pesticides use by a marginal amount to be a part of it. It's a good result overall, though the accreditation body has been criticized for setting the sustainable bar too low for a certification.

Biodynamics

It's impossible to talk about biodynamic farming without mention of Rudolf Steiner, the man behind the founding principles of biodynamics and the spiritual teachings of anthroposophy. There are currently over 800 biodynamically certified vineyards globally, and their numbers are increasing.

Steiner was a bit of a character to say the least, an Austrian occultist, architect and clairvoyant (amongst other things), and the inspiration for the Waldorf schools around the world that practise his spiritual teachings to this very day.

At the age of 9 Steiner claimed to see the spirit of his dead aunt and formed a connection with the spiritual world, linking it to geometry. By 15 he believed himself to have a full understanding of the concept of time, a precursor to clairvoyance in his view. He went on to study at Vienna Technical University, where aged 21 in 1882 he was given the task of editing the works of a German literary marvel Johann Wolfgang von Goethe. Goethe also dabbled in science and spirituality, and it was Goethe's work which formed the backbone for his own belief system.

Many influential agriculturists and landowners were concerned about the new technologies that came out of Fritz Haber's work and their effect on farming, so they sought the help of Steiner. They longed for a systematic approach to farming, one that didn't rely on the new, synthetic chemicals. With some persistence they managed to secure a promise from Steiner to develop said approach. In 1924 Steiner rolled out a series of eight lectures in Kobierzyce, Poland. These lectures set out the framework for what we now call biodynamics, a new and sustainable agriculture that didn't rely on the use of agrochemicals and looks at the farm as a connected single organism. (Interestingly, biodynamics arguably gave rise to organics, when British biodynamic farmer Lord Northbourne popularized the term 'organic farming' published in his manifesto on organic agriculture 'Look to the Land' in 1940.)

Biodynamic producers aim to make their vines strong enough to combat any disease by itself, using nine homeopathic 'preparations' to achieve this. Three of these preparations are sprayed on the vines and six are blended into the compost. Steiner's initial lectures have a strong focus on soil health, and with remarkable insight he spoke of how artificial fertilizers would only destroy the health of soils and their ability to recover health. As zany as he may have appeared, he was bang on the mark with some of his projections. Biodynamics has come in for much criticism since not a lot of it is backed by science and it chooses to place itself in the realm of spirituality. However, many growers swear by the results and simply cannot explain why it works. I hear first hand from winemakers who hold this opinion, farm biodynamically and do gain incredible results. They have faith in the system and don't need to understand the why. One fascinating biodynamic grower and winemaker we work with is Agnieszka Rousseau of Winnica Wieliczka in Poland. Her side hustle outside of winemaking is as a consultant for biodynamics. She's one of only three people within agriculture in Central Eastern Europe on the Goetheanum list of certified biodynamic advisors and she has a background in pharmaceuticals; this, along with her chemical know-how, really comes in handy. Talking with her I've discovered that any scientific evidence behind biodynamics is still a work in progress; while much can be explained now that could not before there's a fair bit we're still in the dark about. I'll do my utmost to cover as much science behind this practice as is currently available.

"Not a lot of biodynamics is backed by science ... but many growers swear by the results and simply cannot explain why it works."

The preparations and biodynamic rules on farming

These are integral to biodynamic practice and clearly differentiate it from organics. Each preparation has a code name ranging from 500–508 which is said to have been a legacy from Nazi Germany. Biodynamics had a very mixed reception at the time; Nazi Germany forbade the practice and creating coded references for the preparations helped German biodynamic farmers practise without unwanted attention.

All nine preparations are composted by leaving them in the soil for 6–12 months stuffed in cow horns or other animal parts such as bladders, intestines or skulls. It's a little tricky to keep the tone of this scientific as we run through the preparations, but we'll do our best whilst noting there is much further holistic Steiner explanation on each that we won't be dwelling on.

500 – horn manure

An unusual and almost cultish approach: cow horns filled with manure from organic or biodynamic cows are buried in the ground. After 6 months of composting, these are dug up, the contents diluted and mixed in a dynamizer and the resulting preparation sprayed over the vineyard soils. A dynamizer is a tool designed especially for biodynamic preparations which allows water to flow down swirling waterways. At the end of each spiral the water is channelled to another receptacle that forces the water to swirl in the opposite direction. It allows the watery contents to mix and plays to Steiner's logic of shapes and geometry.

Steiner describes preparation 500 as an 'extremely concentrated living manure'. From a scientific viewpoint we know the benefits of nitrogen-rich manure and composting would certainly aid the breakdown of raw manure and permit nitrogen to be mixed into a vine-soluble solution. Further to this, animal carcasses house an abundance of microbes, so the use of any already decaying animal part would only serve to aid the degradation of manure into its constituent parts, thus speeding up the availability of nitrogen and other nutrients in the manure. By spraying the final product over the soils we're adding more nitrogen and other nutrients, and introducing a wide array of microbial friends back into the soil mix. Many biodynamic farmers I've spoken with were initially sceptical about this seemingly witch-crafty practice and simply replaced the horns with jars or plastic receptacles which were a little easier to get a hold of. They noted that the cow horns made a positive difference to the type of manure created, and directly benefited the vines.

501 – horn silica

This is prepared in a similar way to the horn manure, but the horn is filled with fine ground quartz before being buried for the summer, dynamized, and sprayed on vine leaves. The school of biodynamics would say this practice aids in photosynthesis, reinforcing the plant immune defence, and in drying up the vineyard atmosphere. From a scientific point of view we might surmise that the quartz addition plays a small part in making the soils paler over time, thus reflecting more photosynthetic light back onto the vines. It should be noted that 501 is used in the most minute of quantities, approximately 4 grams per hectare, so it would take quite some time before any significant changes in soil colour are made, if any at all.

502 – yarrow/503 – chamomile

Yarrow is a common flowering plant from the Asteraceae family and is a rich source of potassium and selenium. Preparation 502 is made by picking the flowers, drying them and then packing them in a deer bladder. The stuffed bladder is then hung to dry before being buried in the soil for the season. It's then dug up and the remains added to compost. Steiner mentions that the deer is a very different animal from the cow, whose antlers point outwards towards the cosmos, and the bladder is a sensitive organ 'almost a copy of the cosmos'. Biodynamicists claim this preparation is good for regulating potassium metabolism in the plant, as well as regulating nitrogen, carbon and sulphur processes.

Preparation 503 is very similar to 502 in the preparation of the flowers, this time chamomile, however here we use cattle intestines instead. It is said to make the soil more nitrogen resistant.

My only note in the science behind these preparations would refer again to animal organs being rich in microbial life, which should in fact aid in turning nutrient-rich yarrow into a form of potassium and selenium that's easier for the vines to take up. There may in this case be a flicker of evidence behind a seemingly nonsensical practice.

504 – stinging nettles

These are said to 'stabilize the nitrogen content of the soil'. It's a simple preparation that requires one to bury the leaves of freshly picked nettles. The use of nettles is reminiscent of the purpose Eric Petiot suggested for them in organic farming. There have been a lot of developments in understanding how living yarrow, chamomile and nettles work to develop mycorrhizal networks and aid in keeping certain pests at bay, although this is when planted in the vineyard and not when applied as a preparation. So perhaps here the real benefit lies with the growth of the plants themselves at the vineyard, as opposed to the ritual of the preparation itself?

505 – oak bark

In this preparation oak bark is grated into the skull of a horse, pig or sheep, buried, dried and added to compost. Steiner comments that oak bark adds calcium to the compost to improve plant defence. Astonishingly, tree bark does have an extremely high calcium content. In the UK there exists the phenomenon of certain grey squirrels stripping back the bark of trees and eating the underlying phloem channels. One theory for this highly unusual behaviour is that it may be done to ameliorate a calcium deficiency at times of need. Female squirrels are most often observed to strip the barks before lactation, and juveniles do the same before growth spurts.

506 – dandelion

Dried dandelion flowers are sewn into the mesentery of a cow – the membrane which holds the intestine in place. Steiner saw the dandelion as a messenger from heaven. Biodynamic farmers state that this is a good stimulant for plant growth, and it concentrates and improves flavours in crops. In gardening and farming alike, dandelions are great to have around as they are an early flowering attractant for pollinators like bees, and have deep roots that do well for increasing the porosity of compacted soils. If planted nearby this would help greatly, however finding a link to the preparation itself may prove quite tricky.

507 – valerian

The pressed valerian flowers are soaked in water, dynamized, then sprayed over compost. Steiner states the valerian will stimulate phosphate activity in the soil. The plants are said to be attractive to pollinators in the field, and the dried roots have been used to treat insomnia since Greek and Roman times. However, any clear evidence to back up Steiner's claims is yet to surface.

508 – horsetail
This common plant is said to be high in sulphur and silica, and to work well as a fungicide. The plant stems are soaked in water, which is then dynamized and sprayed on the vines. An extract of horsetail is also approved for use as a fungicide in the EU and is used widely in other non-biodynamic practices for these purposes too.

It's hard to see the logic behind biodynamics and these preparations may seem wacky, yet it's worth remembering that Steiner created these biodynamic guidelines to combat conventional farming methods and use of newly developed synthetic agrochemicals, practices we now know are detrimental to the environment. Steiner created a system of farming that respects the environment and replenishes the soil. We should at least respect the motivation behind this method of farming. What's most fascinating is that as our understanding of soil and microbiology develops, it corroborates a few of the biodynamic practices we once questioned.

The nine preparations are fundamental to biodynamics, but not the entirety of the system. Steiner's farming methods follow the lunar and astrological cycles with respect to crop management. Planting, cultivation and harvesting depends on the phase of the moon and the zodiacal constellation the moon is passing through. Again, this can be viewed as 'pseudoscience', though it is how numerous ancient cultures around the world came to farm, through a process of trial and error. There are theories on how the gravitational pull of the moon plays a part in promoting root or shoot growth depending on what side of the earth the moon is on relative to the plant. The gravitational forces are there, but so minute we often discard these proposals altogether. My personal take on this is that from a Darwinian perspective, competition is rarely on a level playing field. Those with even the smallest advantage will benefit, tipping the scales in their direction and giving the opportunity for a small fraction of beneficiaries to succeed. In this example, at the right alignment we have the earth's gravitational pull and the addition of the moon's pull. Plant roots display geotropism, which means they find their way by following a gravitational pull. This effect can be simulated in a laboratory: when an artificial form of gravitational acceleration is applied to a plant, its roots follow in that direction. Plant shoots and stems display negative geotropism, which means they grow against the direction of gravity. The minuscule gravitational force of an additional moon's mass may not be doing much to physically draw the roots down, but does provide clearer indications of which direction to travel.

Copper may also be used in biodynamics and is heavily restricted to 3 kilograms per hectare per year over a seven-year average (Demeter standards). In regions with high fungal pressure, exceptions can be applied for increasing this to 4 kilograms per hectare per year over a five-year average. This still sits at being almost twice as strict as the copper restriction applied for organics.

Biodynamics also provides guidance on how to prune effectively to ensure the proper flow of sap and nutrients to the vine, increasing the natural vine defence mechanisms, and methods to avoid soil compaction.

The consumption of alcohol is not encouraged in Steiner's practice of anthroposophy, so it took some time for wine to be included within the remit of biodynamic certification. There are far fewer biodynamic certification bodies than there are organic: Demeter International, Demeter France and Biodyvin are some of the better-known ones.
Like organics, the certifications differ in stringency. Some of the differences between certifications may seem trivial but can completely change how winemakers operate their vineyards as an organism. Demeter for example requires only organic or biodynamic manure to be used. When you look at the practicality of this in secluded farms and vineyards where local cooperation may not be viable, the only option is to keep your own livestock. For most vine growers and winemakers, tending to animals is not where their skills lie. From my experience of sourcing wine for the bars, there are numerous wineries practising biodynamics that go uncertified for these sorts of reasons, it's a stringent and unsurprisingly bureaucratic process that adds great cost to one's wine.

Biodynamic rules on wine production

We can sensibly assume that if biodynamics has gone to such great lengths in the vineyard, this ethos will continue in the winemaking. Use of the following is not permitted for Demeter certified wines:

- Any genetically modified products;
- Addition of ascorbic acid (vitamin C);
- Use of isinglass (sturgeon swim bladder), blood or gelatine;
- Pasteurization.

The following practices are permitted, often with restrictions:

- Adding sugar or grape juice concentrate to increase the alcohol content by a maximum of 1.5% abv.
- Indigenous yeast and *pied de cuve* (a little starter batch of fermenting wine). Bought-in Demeter-certified neutral yeast is permitted only for a stuck fermentation.
- Tartar stabilization by cold stabilization, but only natural tartrate from Demeter or organic wine production are permitted, including potassium as well as bitartrate additions.
- Lactic acid bacteria as a biological acid reduction.
- Preservation with sulphur up to certain levels, approximately 35–40 per cent lower than that of organic restrictions.
- Permitted fining agents include egg white, milk products, bentonite, and casein, pea, potato and wheat protein.
- Only oak wooden barrels are permitted for oaking wine. Oak staves and wood chips are not allowed.

Regenerative farming

'Regenerative farming is like biodynamics, without the religion', quipped Tony Chapman, the viticulturalist and winemaker at The Donham Estate winery in Sonoma, California, one of just two wineries in Sonoma that are Regenerative Organic Certified. Regenerative farming is a science-led farming practice which takes into account the rehabilitation of the soil, animal welfare, carbon sequestering potential and the overall life of the farmer.

It may well be an iteration of what future farming will look like. Thus far it has proven to show no reduction in yields and little damage to the planet, though a significant amount of expertise and knowledge is required on each farm to pull this off successfully.

Core principles of regenerative farming include the use of cover crops, crop rotation, biodiversity, no tilling or minimal soil disturbance, rotational grazing of animals, and use of compost, whilst discarding use of synthetic pesticides and fertilizers. This farming method looks at the information we've discussed in the last two chapters regarding soil health, geology and environment and sets to farm with respect to these principles.

There are just a handful of regenerative farming certifications which include Regenerative Organic Certified (ROC), Land to Market – Ecological Outcomes Verification (EOV), and Certified Regenerative by A Greener World (AGW).

Natural wine

We can keep this brief: legally there is no definition regarding what natural wine is or isn't. Unlike organics, biodynamics and regenerative farming, there is currently no governing body or certification. This means anyone can slap the term 'natural' on a bottle of anything, so it means very little.

From a commercial sense, and from operating a 'low intervention' wine bar myself, the term natural wine is more an ethos than anything else. Most who have the intention of making natural wine, and calling it that, will work holistically and aim to limit the human or chemical intervention in both farming and production. In the vineyard, most will use organic or biodynamic fruit, whether certified or not. In the making of the wine, little will be added or taken away. It is always valuable to remind oneself however that the term is open to abuse.

There are many winemakers who have banded together to create their own groups and set their own restrictions on what natural wine may look like, but of course this is a purely subjective take on the word, and these rules are based on trust since nobody is paid to go around and check! An example of such a group in France is L'Association des Vins Naturels, made up of more than 50 wineries, and Vin Natur in Italy which comprises over 140 members. Even with the natural guidelines set by these groups, the addition of small amounts of sulphur dioxide is mostly permitted.

It is utterly mind-blowing how recently these industry transforming technological advances (like the Haber–Bosch Process) came about, and how quickly our farming changed as a result. Right now we have no obvious nor viable solution for the damage it causes, and on a large scale a reduction in harm is all we can hope for till we find a way.

I'm a firm advocate for hybrids as the future. *Vitis vinifera* is what we know, but it's delicate and requires a lot of care and attention. On the scale and price points we largely push for, its weaknesses can only be overcome by chemical intervention. On a recent visit to one of the Demeter-certified wineries we work with, Vinné sklepy Kutná Hora in the Czech Republic, their lead viticulturist Jakub and I were discussing hybrid grapes and their use of the popular PIWI (Pilzwiderstandsfähige Reben) hybrid Solaris. PIWI varieties are bred for fungal resistance and heightened tolerances to both powdery and downy mildew.* We sell their sparkling Pet-Nat Solaris as our pouring sparkling wine by the glass at both bars, and it's a firm favourite with regulars, at a reasonable price. I was curious about how easy a hybrid Solaris was to grow, and Jakub was adamant that it was one of the easiest varieties he's ever had to work with in that climate. It needed very little, to do very well. A world apart from *Vitis vinifera* which needs endless care and attention and is constantly under fungal attack in those parts. However, Jakub is an ex-Burgundy vineyard hand, and he also admitted that he'd never achieve real excellence from the Solaris when pitted against Burgundian standards. For larger scale, well-priced and smashable wines we shouldn't be afraid to plant more hybrids. They're easier to work with and require fewer chemical corrective measures.

In December 2021 the EU made a huge step forward, modifying their Official Journal to permit resistant varieties in the production of wine within Protected Denominations of Origin (PDO). It allowed use of hybrids within pre-existing regional wine laws should a member state wish to introduce them.

The decision by the EU was as a direct response to climate change, and a move to build the foundations for a more sustainable wine growing industry. Hybrids are still a rarity in wine production, and marketing may be the deciding factor for them. Consumers and wine buyers tend to pick and work with wines they are familiar with. If choosing between a Merlot or a 'Muscat Bailey A' on the shop shelf, most will choose to spend their hard earned money on the thing they know. This choice feeds demand, and growers risk not selling their grapes if they invest in planting hybrids with weird and wonderful yet unfamiliar names.

** All PIWIs are hybrids, but not all hybrids are PIWIs. Many may think hybrids are similar to genetically modified plants but in fact hybrids are simply vines that have been cross pollinated from two different species. GM plants are created in laboratories where their genes can be manipulated to produce a desirable output.*

VINNÉ SKLEPY
KUTNÁ HORA
PET NAT
WINE PUNK
SOLARIS 2022
ČESKÉ ZEMSKÉ
BIODYNAMICKÉ VÍNO

4. The anatomy of a grape

The anatomy of a grape

My first wine training session at Gaucho Restaurants was an eye-opening experience. Fresh out of casual dining, I had been hired as a waiter and entered another world, one of Argentine high epicureanism. Phil Crozier, the director of wine at the time, was very hands-on and led the entire group's wine training. Phil was humble, personable and passionate about sharing his knowledge. These qualities made him ideally suited for gearing up the hundreds of waiters the group employed, most of whom, like me, had little or no fine dining, wine or Argentine experience.

Phil wore some pretty iconic shirts, something I soon realized was a wine thing, and in his first training session he sketched up a grape and spoke a little on what was found within. It was this anatomical overview of the grape that really piqued my interest in wine; it was the first time any part of wine training made any sense. In this chapter we delve a lot deeper into this approach and shed light on how changes inside the grape directly affect the contents of your glass.

For this chapter I would like to take one parameter for granted: the supply of water and nutrients. I'd like us to assume we're working with experienced winemakers who will always aim to keep our vines in a condition where they direct most of their energy to the grape and not to green vigorous growth. This is most often achieved by managing the flow of water to the vine. Whether it's for the bulk production of grapes going into Echo Falls or for smaller runs of Champagne, it is in the winemaker's interests to give their vines the best chance to produce grapes suited to the wine being made, whether that be judged by volume, concentration or some other metric. This is the case for all commercial producers of wine. Neither winemakers nor growers at Echo Falls or Krug Champagne would be in a job without being able to achieve this. For this reason we should assume competent growers are in play.

These days I endeavour to follow in the footsteps of Phil Crozier by doing my own day-one wine training for all new recruits. I call this session 'The Anatomy of a Grape' and it's a great opportunity for new starters to pick my brains, and equally for me to do some brain picking myself to really understand how they each think and learn. One of my favourite questions to pose to new recruits, just to get them in the right scientific mindset, is 'What's the point of fruit?'. We all know the answer, but often think of fruit as something that exists purely for human enjoyment. With some prodding, most budding sommeliers do get to the right answer – fruits are the reproductive vessels for many plants.

At the beginning of their development, fruits exist in a sufficiently inedible state to protect the developing seed. When the seeds are ready, they send a signal for fruit to mature and ripen. This triggers the softening of the flesh and a change of flavours, aromas and colour. Plants have evolved to ripen and attract the attention of animals for the dispersal of their seeds, and not just to satisfy the greedy bellies of humans. This stage of transformation of colour and ripening is referred to in the wine world as 'veraison'.

Where do aromas come from?

To fully wrap our head around how natural aroma molecules are created, we must first understand the living process that makes them, metabolism. Living cells use metabolism as the building process for creating carbon chains. Shorter carbon chains of fewer than five or so carbons are usually light enough to enter the air and can float far enough to reach our noses. Longer chains which often include hydrogen and oxygen tend to be too heavy to get airborne and stay put. This is how the basic 'starter set' of natural volatile aromas is formed: aromas are building blocks made of small and simple carbon chains. The four main chemical families or 'starter-set compounds' of carbon chains are acids, aldehydes, alcohols and hydrocarbons. These aromas can either be metabolized or made volatile by longer carbon chains being broken down into smaller lengths by other processes (such as fermentation). Hence the process of metabolism ties carbon chains together, and other processes like fermentation will break long chains apart into smaller and often volatile airborne carbon chains.

"Shorter carbon chains are light enough to enter the air and float far enough to reach our noses."

The mesocarp – pulp

Let's begin with the juicy part. Most of what's in our wine is formed from the pulp. Of the entire grape, the pulp contains around 50 per cent of the sugar content. It's also where 35 per cent of the acids, roughly 20 per cent of the aromas and between 10 and 20 per cent of phenolics are found. We'll look into phenolics when dissecting the skins.

Acids

Acids are a fundamental feature of wine, affecting the taste, colour, stability and even the potential for oxidation of a wine. In the colloquial language of food, acidity might be viewed as a pejorative word. 'Acidic' doesn't sound like a complimentary way of describing your grandma's spaghetti bolognese, but it is integral to our enjoyment of it. Without acidity, food would be cloying and hard to swallow, and drinks would feel flat, yet still we don't hold 'acidic' with the reverence it rightly deserves.

We can divide acids into two forms: volatile acids and fixed acids. Within a raw grape levels of volatile acids are near negligible, so we will just explore the main fixed acids in this chapter.

Malic acid

Its name stems from the Latin word for apple, *mālum*. A useful way to remember the link between malic acid and green apples is to think of the term 'bad apple', where *mal* in many Latin languages means bad. (This negative connotation between bad and apple stems from the association between apples and the forbidden fruit eaten by Eve in the Garden of Eden. It's no surprise *mal* gets a bad rep.)

Malic acid is the primary acid in many fruit including apricots, blackberries, blueberries, cherries, peaches, sour apples and of course grapes. It's found in abundance in the tart, green almost unripe-tasting Granny Smith style of apple, and it's responsible for giving this apple its unique and recognizable flavour. Malic acid can be tasted in many white wines: it is clearly distinguishable in tart, green appley Sauvignon Blancs and many a Riesling or crisp Chardonnay we will all have sampled. For the vine, malic acid is involved in several processes which are essential for health and sustenance.

The concentration of malic acid in grapes is variety-dependent: Barbera and Sylvaner, for example, have terrifically high levels. In all grapes, malic acid is at its highest just before veraison. As the grapes ripen, malic acid is metabolized and burned up like fuel for the process of respiration, reducing the levels of malic acid in the grape over time. Before veraison, it is sucrose sugars that are metabolized in respiration; only at the beginning of ripening is malic acid used up. This process of respiration is accelerated by heat, so in hotter climates the rate of malic acid decline is much higher. This means in hotter climates you can broadly expect wines of lower acidity compared to the same varieties in colder climates. In this scenario we're considering the total amount of heat applied to the vine – in hotter regions with a high diurnal range (cold nights and hot days), the cold nights will play a part in reducing the rate of malic acid decline and maintaining higher final acidity in the grapes. This then translates to higher acidity in your grape must (crushed grape juice), and more than likely in your final wine.

Tartaric acid

Although it is one of the most important acids in winemaking, tartaric acid is surprisingly rare in plants. Luckily grapes contain an abundance of the stuff! Grapes such as Palomino and most black *Vitis vinifera* grapes in general yield far higher concentrations of tartaric acid. This strong acid is formed in the grape flowers and then relayed to the young berries. Unlike malic acid, these acids are not metabolized via respiration and remain relatively stable during ripening, hence they are affected less by climatic temperature. Tartaric acid concentrations in grapes do decrease during ripening due to dilution caused by the swelling of the grapes with water, but the overall quantity of tartaric acid present remains stable.

Tartaric acid is odourless. When chilled, it precipitates around naturally abundant potassium ions in the wine to form little crystals. These crystals are harmless and will disappear again if warmed. It's these tartaric crystals in wine that led Louis Pasteur in 1848 to question and eventually change our understanding of all living things. The tartaric crystals aren't just repeated lattices, they grow in mirrored patterns. This led him to extend the concept of chirality and reflected symmetry in crystals to that of nature and its life forms, which do the same, for example, the reflected left- and right-hand forms found in most animals. This led later scientists to apply these principles to DNA, spurring the discovery of the double helix. Where would we all be without wine, eh?

If permitted in a given winemaking region, tartaric acid is often used as a powdered agent when adjusting wine to increase acidity (acidification). It has a high degree of microbial stability and strong influence on the total acidity of a wine.

Citric acid

This is the most plentiful acid in citrus fruits such as lemons, limes and grapefruits, and responsible for their particular citrus flavours. Citric acid is found in lesser quantities within the grape compared to malic and tartaric acid. However, it's still another big contributor to the taste and smell of many white wines, in which it is found in varying concentrations dependent on grape variety. Citric acid is also a key component to the Krebs cycle which is used by organisms that respire, so vines need to maintain good stores of it for when respiring the most. And where better than in the grape?

Citric acid is rarely used as an acidifying agent in wine production as it imparts a citrus flavour and smell. It's also not as microbially stable as tartaric acid due to many types of bacteria using citric acid to aid their metabolism. Having lots of this microbial food source lying around will promote the growth of unwanted microbes in your wine.

Ascorbic acid

Commonly known as vitamin C, ascorbic acid is found in young grapes before veraison, but it's very quickly metabolized to near insignificant levels during the ripening process.

"In hotter climates you can broadly expect wines of lower acidity compared to the same varieties in colder climates."

How acidity is measured at the winery

Fixed acidity in grape juice or must is measured by TA (titratable acidity or total acidity – different measures, giving different readings, using the same abbreviation – confusing or what!) or using the pH scale. In winemaking, high acidities are generally safer since most bacteria and fungi are unable to grow and thrive at low pH levels and the risk of bacterial spoilage is reduced. Higher acidities can also denature some of those enzymes that cause browning and oxidization, hence higher acidity grape musts and wine have more stable antioxidant properties.

Tasting acidity

So when tasting how do we know, without lab equipment, if a wine has a high or low acidity? Tartaric acid throws a spanner in the works because it carries very little odour, which means we can't rely on smell to isolate the acidity of our wine. But we don't need to take litmus tests to dinner: our mouths are geared up for this challenge, we just may not realize it yet. Mammals like us automatically produce saliva when potentially dangerous ingredients are introduced to our delicate mouths. Saliva has many functions for digestion; a simple one that's relevant to acids is the act of dilution. Potentially hazardous acids that may damage the soft tissues in our mouths or cause dental erosion are made safer by watering down with the saliva we produce. The higher the acidity of ingredients we pop in, the more we salivate: a simple yet effective acidity assessment system for food and wine.

Sugars

Grape pulp is largely made up of water, followed by sugars then organic acids. The main two sugars found in wine are glucose and fructose. These vary in proportion for each variety, hence their differing sweetness levels. Fructose is twice as sweet as glucose. For a winemaker, knowing the levels of each are important, and knowing which grape varieties tend to have a genetic disposition towards creating one type more than the other is also useful. Chardonnay and Pinot Blanc are considered to be high-fructose varieties, whilst Chenin Blanc and Zinfandel are high-glucose varieties. Both sugars are fermentable, which means they can be used up in alcoholic fermentation. Consequently, the levels of glucose and fructose in grape must will have a direct impact on the alcohol potential of a wine. The sugar content of the grape must is measured in 'Brix', which represents the number of grams of sugar per 100 millilitres of juice.

We cannot smell sweetness since the molecular weights of sugars are far too large to become volatile enough to reach our noses under normal atmospheric conditions. With wine, our perception of the smell of sweetness is just a psychosomatic association with sweet-tasting things. If we smell mangoes or cherries in wine, it may lead us to assume the wine might be sweet like mangoes or cherries, yet in most wines it's highly likely that all the sugars have been consumed in fermentation. A simple experiment to prove this: smell a bowl of sugar. It has no smell, and certainly doesn't smell sweet. This technically means we cannot smell if a wine is sweet, we have to taste it first.

Sugar accumulation within the grape is most rapid during the start of ripening and gently slows towards the end. It's the process of photosynthesis which creates sugars in the grape. The speed of sugar accumulation is directly proportional to the amount of sunlight that the vines are exposed to. Mo' sunshine, mo' sugars – and higher alcohol potentials after fermentation. Photosynthesis takes place comfortably between 18 and 33°C (64–91°F), colder weather or extremes of heat will compromise the vine's photosynthesizing power. Leaves are the photosynthesizing processing centres for the vine, creating sucrose (a sugar molecule made up of both glucose and fructose). The sucrose is then transported to the grapes in a sugary water solution via the phloem (plant channels used for nutrient distribution). This sucrose is dispatched within the grape to cell vacuoles where they are hydrolysed into both fructose and glucose for stable long-term storage. The speed and flow of this sugar solution through the phloem to the grape is controlled by transpiration, therefore heat also plays a part in speeding up the rate of sugar delivery to the grape.

Climate change has a bad habit of altering wine styles in well-established wine regions: when temperatures and sunlight hours soar, so do sugar and alcohol levels. Increased heat also serves to reduce malic acid levels and thus overall acidity in wines from these regions. Wine laws in essence are the parameters for protecting a regional brand, and wines produced under these PDO laws have to fit a certain style expectation. As many wine regions are getting warmer it's becoming increasingly difficult for wine growers to meet these stylistic expectations with rigid PDO rules in place. Bordeaux is a brilliant example of an Old World wine region steeped in history and tradition that has had to adapt due to climate change pressures. In December 2021, the Council of Wine in Bordeaux permitted the use of six new grape varieties in the region's blends. This unprecedented addition to a historical wine region means that it now permits the warm-climate-suited grapes Arinarnoa, Castets (from south-west France), Marselan (used in the Languedoc), Touriga Nacional (the national grape of Portugal), Alvarinho (or Albariño from Portugal/Spain) and Liliorila. This notable change to Bordeaux wine law enables Bordeaux wines to keep their character and style as temperatures and sunlight hours increase. It looks like many more PDOs may be due a revision, and if we are having to change these wine laws around climate change, perhaps we need to completely revisit how they're structured.

Dr Kate Gannon, an LSE Researcher for climate change in the UK wine sector, comments:

The UK has benefited from warming temperatures, and large areas of England and Wales now fall within the ideal climatic range for a range of commercially popular wine grape varieties. Producers have responded and planted commercial vineyards as far north as Yorkshire, particularly focusing on sparkling wines using the classic Champagne grapes Pinot Noir, Chardonnay and Pinot Meunier. But warming is continuing to accelerate and, even in the UK, the best years for growing these varieties may soon be behind us. As the sector has expanded, the UK has been following in the footsteps of its more established Old and New World wine region counterparts and rushing to develop its own set of PDO designations, such as 'English Quality Sparkling Wine' and the more recent 'Sussex' designation. Amidst rapid temperature change and changing viticultural suitability, we need to ask whether this strategy is wise, when these PDOs have the potential to limit the opportunities for producers to keep adapting their production to maximize the opportunities of a changing climate.

Methoxypyrazines

These aromatic heterocyclic compounds can be found inside the skins, stems and pulp. They're responsible for green, grassy, vegetal or herbaceous aromas. More complex mixtures of multiple methoxypyrazines (of up to 20 types or more) can even smell like cocoa or present darker and more nutty aromas. Methoxypyrazines are typically found in the flesh of green capsicum peppers and contribute to the same sorts of smells found in wine. Bordeaux varieties such as Cabernet Franc, Cabernet Sauvignon, Sauvignon Blanc, Merlot and Carménère are genetically predisposed to produce higher amounts of methoxypyrazines compared to most other varieties.

From an evolutionary perspective, the undesirable 'green' odours of high methoxypyrazine content are utilized by plants to deter seed dispersers like birds from consuming the fruit before the seeds are ready. Levels of methoxypyrazines quickly reduce as the grape ripens, and the rate of decline is also accelerated with increased heat and exposure to direct sunlight via photodegradation. Take the real-life scenario of shaded patches of grapes on the vine: they will likely ripen later than the rest since they see less sunlight, and will therefore also retain higher levels of the undesirable green methoxypyrazine smells for longer. It's the perfect natural system to ward off seed dispersers until the grapes are ready to be consumed. Because of this, grapes when harvested too early will create green or vegetal tasting wines. Trace amounts of methoxypyrazines found in wine may be desirable for certain styles, but larger amounts can often create unpleasant wines. Levels can be managed best at the vineyard by canopy management strategies, ordinarily cutting away leaves to let more sunlight in to reach the grape bunches directly. Some insects can also generate extraordinarily high amounts of methoxypyrazines: ladybirds can often be an issue if found in the must and can completely 'green taint' the wine if not removed after harvesting.

Though methoxypyrazines are present in grapes, they are only found in very small quantities, yet have incredibly low aroma detection thresholds – meaning their smell is very easily picked up by people. Humans can detect just a few parts per trillion, hence we're able to pick up even trace amounts by smell. Once formed in the grape, they are very stable compounds so often make it all the way through fermentation to the end of the winemaking process and into your wine.

Hexanol and hexanal

These aldehydes are formed enzymatically from grape fatty acids originating from the pulp and skins of the grape. They're responsible for scents of freshly cut grass and tomato leaf in wines such as Sauvignon Blanc and Grenache. Hexanol is one of the major contributors to the aroma of newly damaged green leaves, serving as an antifungal compound to ward off further damage. In grapes its concentrations are reduced with ripening; like methoxypyrazines they will also contribute to a green and grassy flavour in wines produced from grapes that are harvested earlier. They are both truly wonderful chemicals to let nature know when grapes are good for eating!

The exocarp – skins

The exocarp refers to the two layers of grape skin that protect the pulp against physical damage and pathogen attacks. The red, white, rosé and orange categories of wine all relate to the differing ways the skins of the grape are introduced in the making of wine. Use of the exocarp in winemaking can significantly change a wine's style.

The skins hold around 25 per cent of the entire grape's sugar and are also where 35 per cent of the grape's acids are found. The skins are metabolically active throughout all stages of grape development; around 80 per cent of all grape aromas and up to 30 per cent of the phenolics are formed in the skins. Examples of aromas found in the exocarp are terpenes (see p. 120), norisoprenoid precursors (see p. 143) and thiol precursors (see p. 141) which are often stored as amino acids and sugar conjugates (something bonded with sugar) in the cell vacuoles. When we say 'precursors' we are simply describing the building blocks required for other things to be created at a later stage. In the case of wine, the putting together process is often fermentation.

What are phenolics?

It's a term I've often heard, even at the start of my wine journey. This word was too often muddled into wine explanations, and seemingly accounted for everything and nothing in wine. Sometimes it was said to be responsible for colour or ripeness, but it was also connected to smells. So just what are phenolics and where will we find them in our wine?

Phenolic compounds are secondary metabolites, which means they are not needed for primary plant functions like growth or reproduction but may grant some other selective advantage to the plant. Metabolism has come up a few times over the course of this book: it's simply life's way of taking nutrients and converting them into life-assisting building blocks – like phenolics.

Chemically, phenolics are a family of organic compounds characterized by a hydroxyl group (-OH) bonded to an aromatic hydrocarbon. It's a bit of a mouthful, but examples of phenols we may have heard of in a grape context are tannins, anthocyanins and vanillin. They have reported health benefits which include inhibiting cardiovascular and degenerative diseases, anti-carcinogenic properties, slowing the human ageing process and plenty more. The term 'polyphenols' refers to a group of individual phenols which have come together to form phenolic compounds.

The most renowned examples of polyphenols found in grape skins are those from a subgroup family called the 'flavonoids', which are created before the onset of veraison and develop as the grapes ripen. Here are a few polyphenols that often take the limelight and are pretty significant wine contributors, so pay attention!

Anthocyanins

These flavonoids are water-soluble colour pigments found inside the grape skin cell vacuoles, and there's even a colour in the name: 'cyan'. Depending on the surrounding acidity they will vary from red (low pH), to purple, blue (high pH) or black. This mesmerizing and almost magical effect can be tested when we wash up used red wine glasses. The addition of water reduces the acidity of any leftover red wine residue, increasing the pH value and instantly turning the anthocyanins a shade of blue. In this way, they act like microscopic litmus tests. The change in colour is caused by a slight reformation of the molecular structure of the anthocyanin, which in turn causes the light to reflect differently. Anthocyanins exist in plants as compounds formed from sugars (glycosides) or their non-carbohydrate (aglycone) form, called anthocyanidins. The six most common anthocyanins are cyanidin, pelargonidin, delphinidin, peonidin, petunidin and malvidin. Cyanidin is the most usual type in grapes.

These pigments are produced in grapes during ripening to make them more visible, creating a vibrant contrast against their leaf colour. Highly visible fruit are the most likely to be scooped up and eaten by local wildlife, and more animals finding and eating the vibrant coloured fruit means their seeds are likely to be dispersed more, wider and further. Concentrations of anthocyanins in grape skins are found to decrease in time when there's less direct sun exposure, very similar to how melanin works in our own skin.

Numerous plants metabolize anthocyanins: blueberries, blackberries, strawberries, black beans, red onions and red cabbage all contain high levels. They all display different colours, because of their relative acidity and the type of anthocyanins they contain. When chopping red cabbage, it can get quite messy, often with purple juice left behind on the worktop and plates. If you add a little acidity to the mess in the form of lemon juice you will see the anthocyanins immediately change colour from a bluey purple to red, the exact opposite colour shift from the dirty wine glasses.

Anthocyanins have no smell and very mild astringency. 'Astringency' is a term that refers to a chemical which shrinks or constricts body tissues. It's a word used in wine quite a bit, and always in the context of mouthfeel.

These colour compounds are pretty magical but aren't very stable. Left on their own in your wine, they won't last for very long at all and the colour will quickly fade to brown. One way we can make them a little more stable is through a natural process called glycosylation or acylation (the addition of a carbohydrate or acyl group to the molecule). This is a process which occurs in certain types of grapes and creates many more forms of anthocyanins (more than 20 types). Varieties such as Cabernet Sauvignon may contain all 20 types, whereas Pinot Noir contains no acylated anthocyanins. It's one big reason why the jet-black Pinot Noir grape makes relatively lightly coloured wines, whereas the Cabernet Sauvignon grape may be a little lighter and bluer in skin colour yet creates such dark coloured wines. Another factor at play in this example is to do with the size of the grape, which alters the juice to skin ratio of the wine. Pinot Noir is not only the bigger of the two grapes but also has significantly thinner skins than Cabernet Sauvignon and therefore fewer anthocyanins to contribute even if they are darker.

Sulphur dioxide (SO_2) is often added to food and wine as a preservative. It can unhelpfully bind itself to free anthocyanins and form an anthocyanin–bisulphite compound; this bleaches the colour of free and unbonded anthocyanins, and hence your wine.

Co-fermentation

In addition to the complexities of acylated and non-acylated anthocyanins, there's also the phenomenon of co-fermentation. How does fermenting white grapes and red grapes create a more colour-stable wine than with just red grapes?

Anthocyanins are planar structures (molecules aligned in one plane). This shape gives them the ability to react easily with other planar structures, allowing them to stack up nicely and squeeze water out from between their molecular sheets, like an Oreo cookie being compressed so that all the tasty cream filling escapes. Molecular stacking protects molecules from reacting and combining with water, which makes them far more stable. Co-fermentation also increases colour intensity and shifts the colour of our wine towards the purple spectrum.

Anthocyanins make up one planar side of our Oreo, we just need another planar partner to squeeze out the cream. The planar structures in the mix are referred to as co-pigments in wine production. The wines of Côte-Rôtie, in the Rhône, are famed for their Syrah–Viognier blends, a seemingly odd choice of red and white grapes. Now we know, by co-fermenting the white grapes in with the red, they're adding more co-pigments from pigment-rich white grapes, which will in turn create a darker and more colour-stable wine.

"Co-fermentation increases colour intensity."

Teinturier varieties

These varieties represent a rare family of grapes which have anthocyanins both in the skins and, unusually, in the flesh too. Analysis has shown there are at least 12 teinturier varieties and the anthocyanin patterns found in the skins differ slightly from those found in the flesh. When bitten into they are black inside and out, thus red wines which use these grapes derive their colour from the whole grape. It also means we can't make white wine out of them. Smoke taint in wine is becoming a major issue in some winemaking regions due to the rising incidence of forest fires. Smoke from the fires imparts strong unwanted flavours to smoke-affected grape skins. Use of teinturier grapes such as Alicante Bouschet and Colorino is a nifty means to bypass the need for skins in red wines. Smoke tainted skins can be removed and discarded and the teinturier pulp can then be used to make red wines.

Even with co-ferments (see p. 115), and the acylated anthocyanins, these colour compounds are still not as stable as our winemakers may like them to be. After four years in bottle, most coloured anthocyanins will have disappeared. I know what we're all thinking now: How can that be true given the existence of so many red wines aged over the four-year mark? There is one other way of keeping wine colour even more stable, and we'll touch on this shortly.

People often say that drinking red wine is good for you, and they may be right, to a certain extent. There are numerous well researched health benefits from consuming anthocyanins which include lowering blood pressure and cholesterol, increased blood flow to the brain, blocking DNA mutations which cause cancers and destroying cancer cells. The whole family of flavonoids is chock-full of healthy added value, and a large reason why skin contact wines are said to be better for our health than white wines.

Flavan-3-ols (flavanols)

Flavan-3-ols can be confusing, so I will simplify them to the headlines. Within the flavonoid family of polyphenols exist types of flavonoids called flavan-3-ols (or flavanols, not to be confused with flavonols, for which see page 119). Appropriately, given their complicated name, they are also among the most complex types of flavanoids due to their ability to form long and complex chains. They are important in terms of wine due to particular flavanols called tannins, which you will notice get mentioned a lot on labels and in tastings.

Tannins

Of the many different types of flavanols that exist, the one thing they all have in common is their ability to bind to and precipitate both proteins and amino acids. This binding ability is responsible for their extraordinarily high astringency. The physical binding of the proteins in our saliva is perceived as a drying sensation on our gums and teeth. This explains the very grippy, chalky sensation we feel in our mouths from drinking skin-contact wines.

Precipitation is the action of making a solid soluble in a liquid. In winemaking, tannins are used to make wines clearer after filtering by making leftover proteins, which may otherwise give the wine a cloudy appearance, soluble. The addition of the tannins precipitates the proteins into a liquid form, which then integrates them into the wine.

Tannins have a binding superpower and tend to form long polymer chains; the length of these chains is measured by the degree of polymerization (DP). Tannins found in grape skins have a DP of 20, as opposed to grape seed tannins which only polymerize up to 15 DP. The length of tannins influences how we perceive them in our mouths – longer DP tannins may feel smoother and silkier than more aggressive, shorter-chained tannins. For this reason we have to be incredibly careful not to split the grape seeds in our wine, as the shorter chained tannins in the seed are simply too unpleasant. As a grape ripens on the vine, the DP for tannins in the skins also lengthens. It's worth noting at this stage, as we talk about the DPs of tannins in raw skins and seeds, there are a lot of processes in winemaking that will alter the state of tannins. These include the type of yeast used, the addition of protein-based fining agents and the alcohol levels of the fermenting environment. Tannins are much more soluble in alcohol, so the higher alcohol levels towards the end of a fermentation will extract tannins from the skins at a faster rate.

In nature, tannins exist, like anthocyanins, as a defence mechanism to protect plants and once more assist in signalling when their fruit is ready to eat. Tannins are designed to make plants inedible. They reduce the nutritional benefit of eating plants by binding onto animal digestive proteins in the stomach and deactivating them. This makes the plant less nutritionally advantageous and thus discourages animals from eating them in future. They are found in plant structures like leaves and bark and act as anti-herbivore defence systems. The tannins in the skins of fruit start off at low DPs and soften to higher DPs when ripe to signal when fruit is ready to eat.

Tannins also allow the plant to select which size of animal it would prefer to eat its fruit. Grapes, for example, have thinner skins and fewer tannins than orange peel. In nature, this would make the thinner skinned, lower tannin grapes appealing to a wider range of animals than oranges, which (also being higher off the ground) would tend to attract a larger animal to eat them, i.e. those with large enough mouths and bowels that can handle the additional tannins. Saliva is a clever and slippery animal defence mechanism against tannins. It contains mucin proteins which are sacrificed to tannins and effectively deactivate them before they hit our gut. As tannins bind to our saliva, its lubricating quality is reduced and thus we feel this drying sensation in our mouths.

In the realm of drink, dry is the opposite of sweet and will refer to a drink that has little or no sugar. In most other social circles, dryness is just a lack of water. The

---

"As tannins bind to our saliva, its lubricating quality is reduced and thus we feel a drying sensation in our mouths."

drying sensation of tannins in wine is often confused with the term 'dry' (such as 'a dry white wine') and can be misleading when used imprecisely to describe wines that are high in tannin. Tannins affect neither the water content nor sugar availability in one's wine, hence dry or drying should not be applied to tannins. In the English language we don't have a single word for the effect we experience when we consume high tannin food or drink, and because of that we often struggle to describe what it is we are feeling. This results in all sorts of adjectives being used: 'grippy' is possibly the most succinct, but 'chalky', 'powdery', 'drying', or just making lip-smacking noises to explain the sensation all get their share.

In addition to their astringent characteristics, when tannins are small enough in size they can trigger the bitter receptors on our tongues. They are at their most bitter when the DP is low (peaking around 4 DP), and at their most astringent at 7 DP, with significant reduction in astringency and bitterness as the chains of tannins become longer.

As well as tannins forming long chains, and binding onto proteins, their grabby superpower also allows them to latch onto other components to lengthen their DP. One useful bond they form during fermentation is with our colourful anthocyanins. This double team is referred to as a pigmented polymer. Here, they form an incredibly strong covalent bond which gives the wine superb colour stability. This is the major contributor to the stabilizing of colour in wines aged beyond the four-year mark; it has also been reported that these pigmented polymers can form rapidly post-fermentation during ageing in barrel. The extra tannins from the barrel itself and oxygen from the pores in the wood of the barrel assist in their development. Pigmented polymers are also far more resistant to the bleaching effects of added sulphur dioxide.

Flavonols

Yes, that's correct, flav**o**nols. Another subtly different named polyphenol from the now quite narcissistic presenting family of flavanoids. Flavonols are the precursor building blocks of anthocyanins and have a very similar structure. From a nutritional perspective they are powerful antioxidants and anti-inflammatories.

Flavonols are found in the skins of both red and white grapes, and have a yellow, straw-like colour (which explains the darker yellow colours of some white and orange wines). Like anthocyanins they also function as filters for damaging ultraviolet (UV) light from the sun. Grapes need this defence in place to prevent the potential mutation of seed cells which could sterilize the seeds, rendering the entire biological function of the grape useless.

Like anthocyanins, their production in grape skins is stimulated by UV light exposure, with higher levels of direct sunlight producing grapes with higher flavonol concentrations. Flavonols are key contributors to co-pigment, greatly assisting the formation of pigmented polymers. This means that red wines made from grapes with high exposure to sunlight will retain their colour for much longer, whilst avoiding the bleaching effect of SO_2. This is a crucial factor in predicting how dark certain wine varieties will present in different growing environments. Heat and sunlight will stimulate an increase in the rate of grape metabolism, producing more of the three flavonoids – increasing tannins and colour in your glass of red wine. In order for us to keep the colour stable, co-pigment flavonols are required, hence sunlight is a crucial factor in colour stability in wine. The sunnier the region, the darker our wines should present for longer. This is the case when comparing the same varietal wines from different regions, such as Argentina and France. Argentine Malbecs will almost always be darker because of higher annual sunlight hours and overall intensity of sunlight at higher altitudes.

There's more to skins than phenols, so before we leave the realm of the exocarp we should round up with a summary of the other non-phenol components and aroma precursors also found in the skins.

Terpenes

Terpenes are hydrocarbons (things made of only hydrogen and carbon atoms) derived from the terpenoid family. You may also see them referred to as isoprenoids. Volatile terpenoids evolved in plants as insect-confusing mechanisms to prevent them from being eaten. If we remember back in the soil chapter we looked at geosmin, which is also part of the terpenoid family and created by bacteria to manipulate insects into doing their bidding, attracting or repelling little critters.

The wider family of volatile terpenoids is responsible for many plant-like aromas, ranging from roses to menthol, eucalyptus, ginger and cannabis. Terpenes exist in grapes and wine either as glycosidically bound terpenes, or unbound free volatile terpenes. Bound terpenes are between two and eight times more common than free-form volatiles, and these bound molecules won't make any contribution to the aroma until they're broken up in fermentation. Aromatic grape varieties such as those from the Muscat family, Riesling, Gewürztraminer and Torrontés all have significantly higher free volatile concentrations than non-aromatic varieties like Chardonnay, which is why they smell so potently aromatic.

Sesquiterpene rotundone

Sesquiterpenes are a class of terpenes. Rotundone is the sesquiterpene responsible for spicy, woody black or white pepper aromas in wine. Syrah, Cot (Malbec), Gamay and Grüner Veltliner are notable varieties genetically disposed to containing high quantities of rotundone. Peppercorn seeds also share the same rotundone sesquiterpene in exceedingly high amounts which accounts for their distinct peppery taste and smell.

What's most intriguing is that 20–25 per cent of the general population are anosmic to it, meaning they lack the physical ability to smell the molecule in ordinarily perceivable odour concentrations. Rotundone anosmic individuals will share a completely different experience when tasting wines that have high levels of rotundone. I see this all the time at the bars when obviously spicy and peppery wines are tried and a fair proportion of guests simply cannot detect it. The look of silent inadequacy on their faces is often what outs them, and it's worth remembering that this perception is not a matter of developing one's palate, but an outright 'aroma blindness' for many.

"Volatile terpenoids are responsible for many plant-like aromas, ranging from roses to menthol, eucalyptus, ginger and cannabis."

Rotundone forms within the skins of grapes, where concentrations increase as grapes ripen. While temperatures above 25°C negatively affect rotundone accumulation, light is demonstrated to stimulate its production. If a cool-climate winemaker wants rotundone and peppery qualities in their wines, canopy management techniques, which increase sunlight to grape bunches, are key.

Potassium (K^+)

Just when you thought we'd left the soil ions behind, this little guy turns up in our skins. Potassium is a key nutrient that vines require through their various growth cycles. It's made available to the soil through natural erosion, animal urine and, in conventional farming, via direct chemical addition. The vine utilizes the potassium ions for many single-use jobs. A potassium ion is the key required to unlock the mechanism responsible for the opening and closing of leaf stomata, making stomata rigid or flaccid to regulate water exit out of the vine. The plant needs to store nutrient ions like potassium somewhere so they are ready to be used when needed.

Grape berries are these natural potassium stores, with most of it being held in the skins. But what difference does potassium make to our wine?

Since the reservoir of potassium is within our grape berries, it makes potassium (K^+) ions the most abundant cation in our grape must. We noted earlier that tartaric acid forms crystals which precipitate around potassium ions. Therefore, a high concentration of K^+ in grape must can lead to excess precipitation of free acids, which in turn will reduce the acidity of the wine, increasing the effective pH value. Think of these little K^+ cations as tiny landing pads for free tartaric acid molecules to rest on, allowing them to accumulate and grow as crystals, thus reducing the acidity concentrations of the surrounding liquid wine. This shift in acidity will have a knock-on effect on the colour of the wine because the anthocyanins will tend towards the blue–purple end of the spectrum, and will reduce the overall biological stability of the wine.

In Australia, most vineyards are rich in potassium, which reflects in the high potassium and high pH values, and therefore low acidity, for most Australian grape juice. During vinification, pH manipulation via the addition of tartaric acid is common practice for most Australian wineries.

The endocarp – seed and surrounding pulp

The endocarp contains the seeds and the small layers of tissue surrounding them. It contains just 25 per cent of the grape's sugars and 30 per cent of the acids but a whopping 60 per cent of the phenolics. These phenolics, though plentiful, aren't all of the sort we'd desire in our wine.

Grape seeds are packed with short-chain tannins and bitter grapeseed oil, both unwanted additions to wine production. Historically we've used our feet to stomp and crush grapes rather than using machinery, to avoid splitting the seeds and releasing unwanted bitter components into the wine. Nowadays seeds are removed with a destemmer and grape press early on in wine processing to minimize the risk of these undesirables ruining the wine. An understanding of grape anatomy allows us to decipher how pressing grapes can influence the components extracted into our wine. First presses and free-run juice (as used for many light rosés and high-quality Champagnes) would include mostly the juice from the mesocarp. Further pressing will derive more components from the exocarp and endocarp. This is why first-press rosés and sparkling wines are more subtle and refined styles of wine, yet still retain good levels of alcohol and acidity.

Genetics

On the topic of seeds, genetics play a huge part in predicting the baseline stock levels of all of the above substances in the grapes we plant. Specific genetic attributes are what we define as the core differences between varieties and influence the decisions for wine planting and making. We've discussed how some grapes have higher levels of certain acids, more tannins than others, and even anthocyanins in very odd places: this is all down to genetics, which adds an extra layer of complexity to learning about wine. Having an in-depth understanding of the underlying genetic characteristics of different grape varieties allows a wine producer to better predict how a wine will turn out.

All grape varieties have parents. We refer to a grape variety as a crossing if its two parents are from the same grape species (e.g. both are *Vitis vinifera*) and a hybrid if from two different species (e.g. one parent is *Vitis vinifera* but the other is *Vitis rupestris*). Crossings can occur by the process of natural pollination or through human involvement, and if we like them, we'll put a name to the resulting plant and make a wine out of the fruit. A lot of Balkan varieties were birthed from communism, manifested in vine development as a singularly focused scientific mindset which wanted to create easy-to-grow varieties that produced good quality fruit in large enough volumes to quench the regime's thirst. Exceptional varieties such as the Bulgarian created Rubin, a crossing of Syrah and Nebbiolo, were born from the collective oenological minds of communist scientists who worked tirelessly towards the same vision. These oenologists produced new and magnificent varieties that were geared towards different objectives from their capitalist counterparts but are still producing magnificent wines.

For centuries we've been choosing and cloning our preferred grape varieties, and these choices have shaped the varieties we see in wine today. Let's say that you were growing a healthy-looking Chardonnay vine, and I wanted to plant your Chardonnay in my back garden. If I planted one of your Chardonnay seeds, it would end up growing to be a random 50:50 mix of its parent genes, like our own children. In other words I'd end up with something unidentical to your original plant. It's not exactly what I want, and in the world of wine where stock genes are a prediction tool used for making wine precisely, what we need are exact clones! Thankfully, cloning plants is much simpler than with Dolly the sheep. If you have a garden or houseplants you may even have done it yourself. We can simply take cuttings, propagate, and plant. There are a few ways to clone a vine, but as with roses, cuttings are often the easiest way to multiply your plant stocks. In addition to grafting, it's another manual and time-consuming process, but ensures we keep varietal consistency.

In practice, there are multiple clones of most single varieties, each one with marginal genetic variances from the next. Varieties can mutate randomly at any moment of cell division whilst growing. Some Chardonnay clones for example may retain more acidity, have thicker skins, contain different base levels of terpenes, ripen slightly earlier than others or have no obvious visible differences. These are marginal variances, but for a winemaker and grower, some marked variances may be prized over others. Examples of Pinot Noir clones include Clone 115 which has smaller yields and is better suited to high quality red wine production. Clone 521, on the other hand, has high yields and larger grapes and is thus better suited to sparkling wine production, in which skins aren't often a key component.

Staying with the example of Pinot Noir clones, some of the tiniest genetic mutations have caused such major differences that we have gone on to class them as different varieties altogether, even if their DNA appears the same. Recent DNA testing confirms that Pinot Noir, Pinot Gris and Pinot Blanc grapes are all one minuscule random mutation of the same variety: they are essentially all one variety. The colour mutation that has occurred somewhere in the DNA of the black Pinot Noir grape, the pinkish Pinot Gris and the green Pinot Blanc, is so slight that it's near impossible to see where in the DNA this has happened. They're classified as having identical DNA, yet the impact on the colour of all three grapes is so great that they are classed as three separate varieties. In our defence, prior to DNA testing, we just couldn't have known.

Even more fascinating is the process of 'chimerism': a biological marvel where several genotypes coexist in a single individual plant – in other words, two or more different sets of DNA and tissues are shared across one organism. Pinot Meunier (technically the same variety as Pinots Noir, Gris and Blanc) is a great example. If you compare Pinot Noir and Pinot Meunier in the vineyard they look very different. The Pinot Meunier plant has a silvery sheen from the fine hairs on its leaves, which look almost floury, giving rise to its name – *meunier* comes from the French for miller. Pinot Noir, on the other hand, does not have this furry coating to its leaves.

"For centuries we've been cloning our preferred grape varieties, and these choices have shaped the varieties we see in wine today."

Chimerism starts at cell division, during plant growth. These growth regions of active cell division are called meristems. If there happens to be a considerable random mutation at this division, then one part of the plant will have a different DNA from another. These meristems tend to grow in planes, therefore chimerical growth will occur across large planes of the plant and stick to certain parts of the organism. With Pinot Meunier, two cell layers have clear genetic differences in the vine. The inner layer of the vine is the Pinot Noir genotype, but the outer layer has experienced a mutation in a gene called GAI1 that makes it unresponsive to a plant growth hormone called gibberellin. The inner layer doesn't have this mutation and responds normally to the gibberellin growth hormone. Experiments have been done to separate the alien outer layer and grow a vine from this, and the resulting plant is dwarfed due to unresponsiveness to the hormone. The chimera Pinot Meunier is also a little shorter than Pinot Noir and produces fruit clusters at the end of every nodule on its canes rather than a mix of fruit clusters and tendrils. Another case where until we understood DNA, we simply could not have known.

Clonal selection is of key importance to commercial wine production, and so far we've just discussed the most commonly grown grape species in wine: *Vitis vinifera*. But other species of grapes will also have different genetic elements to them. For example, a large selection of American grape species have a prominent 'foxiness' about their smell. This animal, almost musky, smell is put down to a unique volatile benzenoid called aminoacetophenone that's present in the pulp. Aminoacetophenone is also responsible for the smell of corn tortillas and is found on Asian fruit bats whose caves reek of that same foxy scent.

"Studies have proven little to no marked changes in grape yields or size in relation to vine age."

Old vines

It's widely assumed that when grape cultivars age beyond the 30-year mark, they incrementally produce fewer or smaller grapes over time. These smaller old-vine yields are said to be more concentrated in all of the grape components we have discussed in this chapter. The rationale is that by producing fewer grapes per vine more energy is funnelled into each grape. As the yields reduce, the same annual units of energy are distributed to fewer grapes which means that each grape gets more energy and therefore higher concentrations of its constituent parts.

When this has been studied, however, the results differ. Research on old vine Zinfandel versus young Zinfandel in California, and altogether separate studies on Shiraz in Adelaide have proven little to no marked changes in grape yields or size in relation to vine age. With so many variables involved in year-on-year climatic conditions, it may be something we once presumed before we studied it properly – or perhaps it is just another marketing myth in wine.

I often bring this topic up with winemakers to see what their experience is. Most mention that past the 20- to 30-year mark there is little difference to grape yields as the vine ages. The vines in time do age, and their immune systems do weaken just as ours do, but the grape yields and wine quality of a 30-year-old compared to an 80-year-old vine are fairly consistent. Time and time again I have heard the same story, yet we are taught and upsold old vine wines on the unsubstantiated idea of better quality.

Living things on the skins

There's plenty to be found on the surface of the grapes. These living things from the vineyard can influence the wine in our glass.

Botrytis cinerea

This necrotrophic fungus grows on a variety of fruit, including grapes. Necrotrophic fungi rely on a host plant, killing its cells to support their own growth. Like most fungi, *Botrytis* loves warm, dark, damp and stagnant conditions. When it takes hold of the grape, *Botrytis* will puncture the skins to access the sugar-rich cells within the mesocarp. If permitted to grow, it'll tear the grapes apart, opening the floodgates for other mould and bacteria to get in, and destroying entire harvests. We refer to this style of out-of-control *Botrytis* as grey rot or grey mould. *Botrytis* also produces a polyphenol oxidase enzyme called laccase, which can very quickly oxidize the phenolic compounds of the grape must into quinones that cause discolouration and a change in fruit flavour and aroma. This is a faster and far more damaging process than the oxidative effects which naturally occur in grape must and wine.

Grey rot can be managed in the vineyard through the use of fungicides, or by smart canopy management to allow airflow and sunlight between the grape bunches, deterring the fungus from growing in the first place.

If managed effectively, there's a silver lining, especially if we're in the business of sweet wine. This positively manageable effect goes by the altogether cheerier name of noble rot. These specifically managed conditions are usually in mesoclimates of high moisture and high sunshine. The moisture creates the prime environment for fungal growth, but the early morning sunshine that follows serves to eviscerate the fungus during the day, keeping it at bay and preventing it from causing too much damage. Suitable vineyards tend to be in naturally very sunny areas near rivers or moving bodies of water, where water is kicked up into the air, creating a high humidity environment. Examples include the iconic wine regions of Sauternes on the River Garonne in France, Tokaji on the intersecting rivers of Tiza and Bodrog in Hungary, and large swathes of Germany that abound with rivers.

In these specific locations, *Botrytis* is managed by allowing growth at night and killing it during the day – day in, day out. When destroyed, *Botrytis* leaves behind tiny little perforations from its piercing tendrils on the skins of the grapes. These little holes allow water to evaporate out of the grape, and the grape shrinks a little in the process. Gradually the grapes lose water and become more concentrated in other components such as sugar and acids, all the good bits we want from the fruit. By harnessing *Botrytis* wine growers can produce phenomenally sweet wines from affected fruit. It's one of the pricier sweet wine options available because of the riskiness of the process, the reduced quantity of grape must due to water evaporation and the niche conditions required. Wines with noble rot character tend to be slightly browned from the effects of laccase oxidation and contain oxidative flavours such as mushrooms, honey and dried fruit – all highly valued traits of sweet noble rot wines.

In addition to the laccase enzyme making oxidative style changes, there are other factors at play. The insides of the grape berries are made available to the elements once they have been perforated by the pesky fungus. These holes become small entrances for other bacteria and fungi to get inside and feast on the sugars within. Common noble rot related guests include *Acetobacter* bacteria, *Alternaria alternata* fungus, *Epicoccum nigrum* fungus and various yeasts. This assortment of foreign micro-organisms starts to ferment the grapes alive. These microbes form many flavour precursors that will come about during fermentation, and trigger enzymatic processes which increase the amount of glycerol produced (thus thickening the wine). Typical aroma precursors created through the action of these organisms include vinegary (from *Acetobacter* bacteria), tropical fruit (from sotolon – discussed in the next chapter), and almond-like aroma compounds.

Wild yeast

If you've ever seen grapes growing in a vineyard, you may have noticed that they don't quite look the same as those found in the shops: they tend to be coated with a waxy white film and a powdery white substance. The waxy white film is called waxy bloom and is a multifunctional secretion (oleanolic acid) which acts to control and reduce the rate of water transpiration from the grape. Oleanolic acid also functions as a chemical attractor for certain European moth species and acts as a natural fungicide to work against the progress of damaging fungi like *Botrytis cinerea*.

The powdery stuff that sticks to waxy bloom is made up of a collection of yeasts, bacteria and other tiny things that happen to get stuck to the bloom and grow little colonies. Numerous types of yeasts and bacteria grow in vineyard soils and can be transported to the vines and stickier grape skins by either wind or insects – not all of these microbes are useful for alcoholic fermentation. Strains of yeast found on the skins include *Kloeckera apiculata*, *Metschnikowia pulcherrima*, species of *Brettanomyces* and several species in the *Candida* and *Pichia* genera. If left to do their thing, these wild yeasts may ferment the sugars in crushed fruit into alcohol. If permitted to ferment during winemaking, they'll create different aromas than we would normally expect from our primary fermentation yeast, *Saccharomyces cerevisiae*. *Kloeckera apiculata* yeasts, for example, are capable of producing very high amounts of ethyl acetate (which has a nail-polish remover smell) and acetaldehyde (with a strong fruity/green apple aroma). These wild yeasts tend to have a very low alcohol and sulphur toxicity tolerance, and at around the 5% abv mark most will be wiped out. In most modern winemaking these wild yeasts are culled by the addition of sulphur dioxide at the start of fermentation, leaving the alcohol and sulphur dioxide-tolerant *Saccharomyces cerevisiae* to continue alcoholic fermentation on its own.

Remarkably, *Saccharomyces cerevisiae* is a minority yeast on the skins of grapes, it's a little too cumbersome to travel easily by air and struggles to settle on the grape skins once there. It does however find a home more easily on shrivelled grapes, where it can nestle nicely in the crevices. Most *Saccharomyces cerevisiae* will come from the yeast-heavy populations within the winery or fermenting room itself, unless added in a powdered, commercially produced form.

The melting pot of different yeasts and bacteria in fermentation will contribute to the final flavour of the wine. If non-*Saccharomyces* micro-organisms are allowed to grow, this will also contribute enormously to wine style. Knowing the microbial communities found on grape skins vary widely from place to place, it leads us to question why we don't consider these micro-organisms as a part of our definition of terroir.

5. Microbes

Microbes

These near invisible little living wonders are responsible for all life above water. They allow us to thrive on this planet; they are the engines of life that feed the mechanisms which permit us all to live. Microbes have always been undervalued in our society, and in the realm of consumer level wine it is much the same. We never mention them within ideals of terroir or see them detailed on the back of a label, even though they can often make more of a difference to a wine than place or climate.

In trying to explain why two wines from the same variety but from adjacent vineyard plots can taste so different, wine experts nearly always seize upon geology. This has been the case in well-established wine regions all over the world, yet we rarely turn our thoughts to the different microbial communities that populate one square metre of soil compared to the next. Research on forests displays clear evidence of microbial microclimates. The findings do indicate a link to geology, in that certain types of microbes prefer some geological formations over others, and as you may have intuited, the smoothness of the material is often a deciding factor. In general, microbes aren't fond of smooth surfaces as they struggle to stick. This forest related research can be extrapolated to vineyards and wineries, but to understand its implications we'll need to better understand the effect of microbes in wine production. Where better to begin than with the star of the show, *Saccharomyces cerevisiae*, also known as brewer's or baker's yeast.

Saccharomyces cerevisiae

We've marvelled at the properties of yeast since the early eighteenth century. It made an appearance in Samuel Johnson's 1755 *Dictionary*, where it was described as 'the ferment put into drink to make it work, and into bread, to lighten and swell it'. Back then we perceived yeast to act more like a chemical catalyser than a living biological entity. It was our old friend Louis Pasteur, who triumphed in the discovery of fermentation as a biological process rather than a chemical or a mechanical one. Pasteur's work was later developed by German chemist Eduard Buchner, who received the 1907 Nobel Prize in Chemistry for working out that it was the bits inside of yeast that did all of the fermenting, and not necessarily the living yeast that was needed for fermentation to occur. He discovered this by simply removing the outer membranes of *Saccharomyces cerevisiae* and using just their innards; surprisingly the yeast insides continued to cause fermentation.

The name *Saccharomyces* derives from the Latinized Greek terms for sugar and fungus, *saccharon* and *myces*, while *cerevisiae* is a Latin-derived species epithet meaning 'of beer'. As well as beer, this aptly named little organism is responsible for the production of bread and wine. The yeast has a crafty mechanism for defending itself and its sugar resources from nearby competitors: it secretes a toxic and psychoactive alcohol. It's an evolutionary trait which it's had to develop a resistance to in order not to poison itself in the process.

In a sugar-rich environment, *Saccharomyces* will continue to consume this source of energy and create ethanol alcohol as a by-product to keep competitors at bay. When its environmental sugar levels eventually deplete these defences are no longer needed. In environments where there is access to oxygen, *Saccharomyces* will start to metabolize the ethanol it created earlier, receiving additional energy whilst metabolizing its defence system. These are highly adaptive little yeasts, and adaptation has made them very successful.

Many other microbes have built up these sorts of microbial defences. A well-known one is the *Penicillium* mould which secretes an antibiotic that breaks down the cell walls of nearby bacteria. We have harnessed its antibiotic properties for our own human purposes in the form of penicillin. Acetic acid bacteria also have a clever mechanism: these acid-tolerant bacteria secrete acetic acid (vinegar to you and me) to create a harshly acidic surrounding environment for their neighbours. The microbial world is a competitive one, and use of chemical warfare assists in gaining the upper hand.

Our main player, *Saccharomyces cerevisiae,* functions differently when working aerobically (with free oxygen present) than it does when working anaerobically (without free oxygen). Traditional wine fermentation is considered to be a near anaerobic process since the yeast will sink below the liquid when wetted with the grape must, and therefore won't have much free oxygen available.

The nutrient content of the grapes we grow influences the nutrient content of the must, which in turn influences the performance of the yeast. It's crazy to think that most of the smells and tastes we get from wine are produced by *Saccharomyces*. Yet its ability to metabolize and create the flavours and aromas we prize so much is heavily swayed by its nutritional state during fermentation. Nitrogen is key to maintaining healthy yeast nutrition; we need it in a form that the yeast can utilize, called yeast assimilable nitrogen (YAN). Vitamins like thiamine and minerals such as magnesium and zinc are also essential for yeast activity. If soils prove to have poor fertility and are nitrogen-poor, this can cause issues with fermentations getting stuck or not even starting, or the yeasts creating bad aromas and tastes. YAN is consumed by *Saccharomyces* as it enzymatically breaks down sugars, which means that the more available sugar there is in the must, the more YAN we need. Broadly speaking, to ferment 1 gram per litre of sugar, *Saccharomyces* will need 1 milligram per litre of YAN. Think of YAN as the oil which keeps the enzymatic engines within yeasts turning smoothly.

Very often in conventional farming, nature's in-built nitrogen-supplying sources are destroyed by the addition of fungicides, so nitrogen has to be applied to the soil in the form of synthetic fertilizers. In addition, wineries will tend to add YAN during fermentation to help the yeast along. We can see that the link between terroir and wine characteristics grows weaker as our technology advances. Before we had the ability to add nitrogen to soils or YAN to fermentation, soils would have had to generate enough nitrogen on their own or we'd help them along a little with organic manure of some sort. In the isolated case of nitrogen, there is a clear and measurable link between soil, wine quality and final wine taste due to the working conditions of our yeasts. However, with the addition of nitrogen at both the growing and the winemaking stage, we sever this connection with the land but ensure that vines and yeasts get what they need to perform at their best.

"The link between terroir and wine characteristics grows weaker as our technology advances."

How does *Saccharomyces* make alcohol?

Yeasts and people aren't too dissimilar in the way we work: at the most basic level all life needs energy. Animals like us get energy by consuming food in the form of sugars and breathing in oxygen. We then convert these inputs to a better, more liquid form of energy currency called adenosine triphosphate (ATP), and exhale out the waste products as water and carbon dioxide. ATP is found in all life forms: it's the universal currency of energy for all living things. The energy equation for humans presents as follows:

$C_6H_{12}O_6$ (1 x glucose) + $6O_2$ (6 x oxygen) → $6H_2O$ (6 x water) + $6CO_2$ (6 x carbon dioxide)

All animals require the key component of oxygen as an input. Think of oxygen as a catalyst that enables us to break down the sugars in our food in the most efficient way: it's the fuel we need to burn food economically and completely. We start with six-carboned sugar glucose, and each bond is broken down into six individual carbon units by use of oxygen and exported as CO_2. With each carbon bond broken, energy in the form of ATP is released. Because of the oxygen catalyst we're able to break down each of the six carbons from sugar and derive the most amount of available energy from each unit of sugar.

Oxygen is the key here, and as we well know, without it we wouldn't be around for very long. I liken this process of respiration to the workings of a car engine. Oxygen provides the spark, and regardless of how much petrol you have in the tank, if there's no spark to ignite it, you're not getting far.

Yeasts and certain other microbes are less dependent on oxygen than we are, and can survive in extreme low- or no-oxygen environments. This anaerobic lifestyle may not be the most efficient form of ATP creation, but it's better than the alternative.

Just like the fuel analogy, with lots of oxygen our yeast will thrive, but as you close in the area and remove the oxygen, the burning fire will weaken and eventually put itself out. In the absence of oxygen *Saccharomyces cerevisiae* can switch over to an anaerobic pathway and still create ATP: it's less efficient but leaves behind a remarkably useful by-product.

In anaerobic conditions we begin with the same glucose input as we did with oxygen-breathing animals. Without the presence of oxygen, the yeast is unable to burn down all six carbon links in the sugar. It does its very best to break down as much as it can on its own and this effort can be represented in the equation below:

$$C_6H_{12}O_6 \text{ (1 x glucose)} \rightarrow 2C_2H_5OH \text{ (2 x ethanol)} + 2CO_2 \text{ (2 x carbon dioxide)}$$

Under oxygen-restricted conditions, the yeast simply does not have the catalytic energy to break down the sugars in its food completely. Ethanol and carbon dioxide are the inefficient – unbroken – remains of our glucose molecule. Unlike the aerobic energy equation where all six carbon chains from glucose are broken down, we've only broken down four chains, two ethanols and two carbon dioxides. It's far less efficient since there are still two more carbon chains left unbroken within the ethanol. It's not a fully satisfactory reaction for the yeasts either, as they need to consume higher amounts of glucose to produce the necessary ATP to keep them going. The more sugar our yeasts consume in this environment, the more alcohol is produced. Unbroken carbons still exist within the structure of the ethanol, but fortunately for the yeast, a useful toxic alcohol is formed in the process. This ethanol still contains unbroken carbon chains that are full of energy, and as we know, the yeasts can only break this down if oxygen becomes available.

Unlike cooking, which relies on heat to bring about chemical change, fermentation relies on yeasts and the enzymes within them to make chemical changes to our food. The word fermentation derives from the Latin word *fervere*, which means 'to boil'. An observation no doubt made after seeing fermentation in full swing. Along with the ethanol by-product, traces of other alcohols, short-chain acids and ethanol–acid esters are created during fermentation. Most living things produce some of these by-products whilst metabolizing, yet nowhere near the abundance that *Saccharomyces cerevisiae* creates. Whilst *Saccharomyces* works away at eating sugars, its enzymes will work to combine and tie together nearby molecules; alcohols which are linked in this way with shorter chain acids (one of the flavour precursors in the grape) will make esters.

Esters

Volatile esters are responsible for the characteristically ripe fruit or floral aromas in wine and beer. They're hybrid molecules made of an alcohol and an acid molecule sewn together by fermentation. Fruits are the primary ester-production units found in nature, hence they smell fruity – a bit of a chicken and egg situation. In fact, there's evidence to suggest that microbes first developed the aromas of fruit and flowers for their own advantage well before the plants we associate them with started developing them. During fermentation we introduce an even wider range of alcohols and acids to the mix, and with yeast there to catalyse the matchmaking, more esters are created. Ester smells include the strawberry-like methyl butyrate, tart apple butyl acetate, passion fruit hexyl hexanoate, and the banana smelling isoamyl acetate that's found in certain wine styles such as Beaujolais Nouveau.

Esters can also be undone and regress to their parent components – this generally occurs when fruit is overripe, or wrapped tightly in storage. Under these circumstances fruits begin to develop vinegary or cheesy notes due to the esters falling apart into their base alcohol and acid components.

It may sound bizarre that a fungus-like yeast would begin to create such fruity and floral aromas in the first place, but, characteristically, nature has a plan. Like fruit and flowers in the wild, ester aromas are produced to lure animals in. Just like plants, *Saccharomyces* would also benefit from a fly or two to buzz around and transport it a little further afield. It's been documented that smaller insects like fruit or vinegar flies are not attracted to aromas of fruits themselves, but to resident yeast alcohols and esters found on the fruits. Like *Saccharomyces*, these flies have also developed a tolerance to toxic ethanol, and feed on the nutritious yeasts, often laying their eggs in high yeast population locations so that fly larvae can feed on them once hatched. This symbiosis of yeast and fly ensures an easy meal for the fly, and by sacrificing a few for the greater good, remaining survivor yeasts are transported far and wide.

Lactones produced in fermentation are also technically esters. They are formed from the same building blocks as esters but, chemically, create different shapes. Esters form straight chains, while lactones bend back on themselves in ring shapes with a tiny tail at the end. They lend wines either particular 'fatty' qualities, such as milk and coconut, or stone fruit attributes like peach and apricot. We will look more closely at lactones in the context of oak ageing (see p. 176).

Esters are a great example of how two flavour precursors can combine to make a potent chemical that changes the flavour of our wine. Understanding the key role of flavour precursors in the grape will give us a better understanding of what further complex aromas can come about from fermentation. Here are a few which contribute most to our sensory experience of wine:

Thiols

Also known as mercaptans, thiols are similar in chemical structure to alcohols. Inside the chemical structure of a thiol, a sulphur molecule takes the place of the oxygen molecule that would normally exist in an alcohol. Thiols are formed by yeasts during fermentation – you guessed it – from an assortment of aroma precursors. Thiols are responsible for a variety of exotic and animal aromas in wine, ranging from tropical-fruit-style aromas of grapefruit, guava or passion fruit to notes of blackcurrant or cat-pee. They are synonymous with the tasting notes of certain New World styles of Sauvignon Blanc, which are dependent on thiols to deliver their distinctive tropical fruit aroma.

Research has shown that thiol precursor biosynthesis is closely associated with the production of certain grape amino acids and their by-products. The development of these amino acids is related to the nitrogen status of the vine, hence nitrogen and sulphur supplementation in the vineyard can enhance thiol precursors in berries.

In trials with white *Vitis vinifera* varieties, nitrogen and sulphur supplementation during veraison proved to increase thiol concentration in wines by up to five times, depending on the variety and growing conditions. Another way in which vineyard management can 'fix' the characteristics of our wine.

Strawberry furanones

These volatile compounds are formed from sugar molecules, and sugars as we know are sweet but not aromatically volatile. Through enzymatic processes these non-volatile sugar molecules can be rearranged into volatile furanone molecules. Furanones are widespread in nature, and responsible for the smells of strawberries, raspberries, pineapples, mangoes and blackberries. These smells are suggestive of sweet fruits, but in high concentrations furanones can be responsible for the caramel-like smells we would expect from browned sugar. This specific caramel-like smell also occurs in cooking when sugars are exposed to higher heat; chefs refer to this as the Maillard reaction and it is the process behind a perfectly seared steak, browned bread and toasted marshmallows, among other things.

Sotolon has exactly the same chemical formula as a strawberry furanone, but is formed from an amino acid rather than a sugar. It is the major volatile aroma in fenugreek seeds and also shows up in smaller concentrations in pineapples. It's a spicy aroma reminiscent of curries, often used as an additive in the production of maple syrup imitations.

"In high concentrations furanones can be responsible for the caramel-like smells we would expect from browned sugar."

Both furanone and sotolon occur naturally in some grape species, with higher concentrations found in hybrid varieties and non-*vinifera* species of grapes. Research indicates that the formation of both is heavily based on the initial sugar content of the must and has a direct positive correlation with oak ageing, a lengthy maturation time and fruits affected by noble rot. The concentration of furanones is also directly related to grape variety, with red grapes containing more than white varieties. Of the white varieties, Chardonnay contains the highest quantities. Furanones play a huge role in forming the final aroma profiles of red varieties such as Primitivo and Refosco.

Norisoprenoids

Norisoprenoids develop mainly in grape skins. They originate from large carotenoid molecules – such as beta-carotene – that are also found in carrots, spinach and broccoli. They accumulate during veraison then break down into smaller compounds as the grapes reach maturity. Within the grape, these compounds are bound to sugars and are therefore non-volatile and aromatically inactive. During fermentation and even as wines age in bottle, they are released from their sugars via hydrolysis and mature into aromatic volatile norisoprenoids. These aromatic compounds are key to forming the varietal character of many wines, including those of prized Rieslings with the two main norisoprenoids being beta-damascenone and 1,1,6-trimethyl-1,2-dihydronaphthalene (which you'll be relieved to know is usually shortened to TDN).

Beta-damascenone is a norisoprenoid whose aromas contribute to smells of berries, rose, and can be a little woody also. It is formed primarily during fermentation.

TDN gives a very distinctive petroleum or kerosene smell, one which is synonymous with good quality aged Rieslings. These norisoprenoids are formed whilst the wine ages and are at their highest concentrations with prolonged ageing.

Carotenoids develop in higher quantities in grapes exposed to direct sunlight in the vineyard. These compounds are involved in photosynthesis and UV-protection for vine leaves and grape skins; they harvest excess light energy and convert it to glucose in a similar way to chlorophylls. Because of this, wines derived from sunnier spots will have higher levels of beta-damascenone and develop higher TDN concentrations either during fermentation or whilst ageing.

Norisoprenoids are detectable by our sensory thresholds at super-low levels. They begin to affect a wine's character at concentrations between 2 and 10 micrograms per litre, although levels of TDN in well-aged Rieslings have been recorded to reach significantly higher levels of more than 50 micrograms per litre.

Terpenes, again?

When we last discussed terpenes, we were in the realm of the exocarp in aromatic grapes such as Muscat and Gewürztraminer, where most of them were glycosidically bound. Although present in the skins they are bound to a heavy sugar molecule and therefore are not volatile enough to be smelled. But *Saccharomyces* wants nothing more than to eat that sugar molecule. Once it gets to work fermenting it cleaves the sugar molecule clean off, freeing up a terpene or two in the process. Linalool, geraniol, nerol and limonene are the most common terpene contributors to our wine experience.

Linalool

This terpene is present in almost 75 per cent of flowers and is responsible for their floral and woody aromas. It is also present in papaya fruit, accounting for the fruit's unique floweriness, and is used as an aroma additive for many household cleaning products and manufactured foods. Though wonderful in aroma, its main function in nature is as a powerful chemical defence against insects such as ticks, so it is also used as an active ingredient in many pet treatments for ticks, fleas and other biting insects.

Geraniol and nerol

These terpenes have a mirrored and identical structure to one another; they fall under the collective term 'citral'. As the name implies, citrals provide the lemony aromas to lemons, limes, oranges and many other aromatic herbs, such as lemongrass and coriander.

Limonene

Like geraniol and neral, limonene imparts very lemony and citrus-peel-like smells. All three of these lemon-like terpenes are found in highly aromatic varietal wines such as Rieslings, Pinot Grigios and Sauvignon Blancs.

Because most glycosidically bound terpenes are found in the grape skins, the terpene content in wine increases proportionally with the length of time spent macerating on skins. Of the four terpenes above, proportions of linalool are often much larger than the other three.

Glycerol

Glycerol is derived from the sugar in grapes and is the most abundant by-product of yeast fermentation after ethanol and carbon dioxide. It is the third largest component of wine after water and alcohol (in dry wines), and it contributes to the smoothness and texture of wine, giving the perception of fullness and roundness.

Saccharomyces cerevisiae produces glycerol whilst metabolizing, accumulating in the cell membrane of the yeast as a measure to balance external osmotic environmental pressure. That's a bit of a mouthful, so let me explain. Osmotic stress in cells occurs when there's a sudden concentration imbalance in the solution around them, which may cause water to rapidly flow into or out of the cell. This is not an enjoyable experience for our *Saccharomyces,* so when faced with osmotic stress, yeasts produce and accumulate thick glycerol in their cell membranes. This serves to regulate external environmental concentration changes, whilst preventing dehydration. Glycerol is like a gloopy buffer that removes the shock caused by the changing external environment.

Glycerol production is accelerated in higher osmotic stress environments, which occur in wines made from botrytized fruit and in sweet wines, where extreme sugar concentrations can be pretty stressful on a little yeast. Extreme low oxygen environments like carbonic maceration (which we'll discuss in the next chapter) will also apply stress to our yeast, creating more glycerol in the process.

"Glycerol gives the perception of fullness and roundness."

Leggy wines

I'm often asked by tasters what the transparent streaks slowly rippling down their wine glasses are. Many misconceive them as an indication of good quality wine.

In fact, the legs of a wine are simply the more viscous parts of a wine, the thick, sticky bits of liquid that are a little slower to move than the rest of the wine. Viscous fluids have strong cohesive forces; their molecules struggle to slide past each other which means it takes them a while to get places. In our wine, viscosity can come from three components: sugar, alcohol or glycerol.

When we see thicker legs of wine unzipping their way down our glasses, this simply means the wine has more of some, or all three, of these components. By using our deductive powers we can make a good guess as to which ones. If it's a sweet wine we'd expect it to be mostly sugar, if it's a dry and high alcohol wine it's likely to be largely alcohol, and if the wine is dry and low in alcohol we can be pretty sure that it's mainly glycerol. In all three cases, higher sugar, alcohol or glycerol contents have never been a quality indicator, which means we can get great wines, with or without the legs.

"The legs of a wine are simply the more viscous parts ... thick, sticky bits of liquid that are a little slower to move than the rest of the wine."

Flor

Flor is remarkable stuff, it's the collective term for the four strains of *Saccharomyces cerevisiae* which have adapted their cell walls to puff out and enable them to float during fermentation. These strains of yeast are utilized extensively in some Sherries, such as fino and manzanilla, and other forms of flor-aged wines, such as the white wines from Jura known as *vins jaunes*. These specially adapted floating flor yeasts are often found on the wine equipment and in the cellars from regions which regularly make flor-aged wines – it's another chicken and egg scenario. Since their base populations are higher in areas where flor-style wine production takes place, they help inoculate more ferments within that space, becoming the dominant yeast in those cellars. Flor's adaptation is a crafty one. By allowing *Saccharomyces* to float, the yeasts form a biofilm on top of the wine. They now have access to previously unattainable oxygen from the air above the wine. Oxygen availability takes our yeasts out of the inefficient anaerobic cycle we touched on earlier and allows them to metabolize more efficiently, enabling them to break down elements they wouldn't have been able to anaerobically. Remember the ethanol which still has that unbroken carbon chain, full of potential energy? Well, now the presence of oxygen allows our flor to metabolize and break it down. Suddenly these clever floating *Saccharomyces* yeasts can start to break down other carbon sources like glycerol too, producing acetaldehyde and sotolon as a by-product in the process.

Sotolon production by flor is faster and more efficient than with traditional *Saccharomyces*; this gives biologically aged flor-fermented wines a more pronounced nutty, spicy, fenugreek and curried aroma. Acetaldehyde is a direct result of alcohol being broken down by the flor and then oxidizing; it carries strong flavours of green, overripe or bruised apples. The combination of spice and apple aromas is closely associated with flor-aged wines.

Flor strains of *Saccharomyces* have adapted a higher tolerance to oxidative stresses due to the nature of their adaptation, and by the process of building a biofilm on top of the wine they also protect the wine from oxidizing. One thing to note about the character of wines they create is that they consume a lot of the more viscous components of a wine. Ethanol, glycerol and sugars can all be eaten up and this means flor-induced wines will tend to be thinner in mouthfeel. They will also lead to marginal reductions in alcohol – sometimes offset at the winery by natural water evaporation during ageing which can concentrate alcohol levels.

Yeast autolysis

The term autolysis refers to the breakdown of a cell by its own enzymes. It's a form of self-destruction and in the case of *Saccharomyces* its enzymes digest it upon death. There are many instances that can trigger a *Saccharomyces* yeast to want to do this to itself, but the most common scenario is when a yeast is overly stressed – be it from ethanol toxicity, nutrient deficiency, older cells weakening and giving way, or sudden shocks in its environment such as rapid cooling or warming. The term lees is often used in winemaking to describe the deposits of dead yeast and other precipitates that settle at the bottom of a fermentation vessel as a result of autolysis.

The first step in yeast autolysis is the degradation of the cell from the inside out. Here enzymes like protease will begin to break down proteins within the cell membrane. The second step occurs after the proteases have managed to puncture through the cell walls. This means that the cell membranes are now porous and allow a mixture of degraded cellular components to leak out of the perforations and into the surrounding wine. Many aromatic compounds are released during yeast autolysis including esters, terpenes, higher alcohols and furanones, as well as other non-aromatic compounds such as polysaccharides and fatty acids. This spillage of the yeast guts is most commonly known to add yeasty and bready flavours to wine, as well as giving it more texture and mouthfeel. It's these physical bits of yeast guts that we feel and that add more volume to wines made in an autolytic style, like Champagne, Franciacorta and Crémant, which all utilize the lees. We will discuss techniques that can be employed to further these autolytic flavours and textures in the next chapter.

Spontaneous fermentation

At the end of the last chapter we mentioned the wild yeast populations which exist on the skins of grapes, yet for larger-scale wine production it is usually commercially grown and cultured yeasts that are used in fermentation. These commercial yeasts are specific strains of *Saccharomyces cerevisiae,* added to the grape must early on to dominate fermentation. They're predictable and enable winemakers to create a more consistent final product, perfect for companies producing large amounts of wine that need to retain a uniform character. During this process other wild yeasts are outcompeted from the offset since commercial yeasts are selected to take over the mix. To further tip the balance in favour of *Saccharomyces,* sulphur dioxide can be added at the start of fermentation to give the industrially grown *Saccharomyces* strains a head start against their more sulphur-sensitive opponents.

Indigenous or wild yeasts are often too unpredictable for large-scale commercial wine production. In wild ferments we have no idea how or if our *Saccharomyces* yeasts will gain a competitive advantage amongst the other microbes that are fighting for their survival in the fermentation mix. Spontaneous or wild fermentation is the term coined for a 'natural' fermentation without the use of cultured yeast. Due to the scarcity of *Saccharomyces* on the skins of the grapes, these ferments take longer to get going because our desired yeast begins as a minority microbe. This introduces a higher element of risk of spoilage if other undesirable microbes take the competitive lead.

A microbial battle for survival

Wild yeast strains such as *Hanseniaspora* and *Klockera* naturally take the lead at the beginning of wild ferments as they're the most populous on the skins of grapes. They have higher enzymatic processing power than cultured yeasts and result in more sensory complexity since they lead to the formation of different esters, alcohols and terpenes. Research from the last decade has also highlighted that some key compounds released during *Hanseniaspora* and *Klockera* fermentation, such as pyruvic acid and acetaldehyde, are precursors for the formation of stable anthocyanin pigments. This will positively affect the colour stability of the wine at very early stages of fermentation.

Hanseniaspora and *Klockera* yeasts have far lower alcohol toxicity thresholds than other yeasts in fermentation so will lose their battle as alcohol levels in the ferment steadily increase. At this point, they are most likely to be replaced by various types of *Candida* wild yeast strains. *Candida* typically grows as a chalky-white film on the surface of a ferment, similar in appearance to flor, and because of its access to oxygen will produce acetic acid and other oxidized end-products.

When the ferment reaches 4–6% abv most wild yeast species are no longer able to survive, and it's here that alcohol-tolerant yeasts like *Saccharomyces cerevisiae* will come to the fore. Wild yeasts tend to be more sensitive to sulphur dioxide and higher temperatures, so if *Saccharomyces cerevisiae* struggles to get a hold, warmer fermentation temperatures and the addition of sulphur dioxide can help it move along.

Many winemakers wishing to ferment their wines naturally will first make the equivalent of a baker's starter batch. We call this a *pied de cuve* and it's often made from a small batch of grapes picked just before harvesting. This small and isolated batch will ferment independently, passing though the stages described above until it reaches a point where the alcohol levels are high enough for *Saccharomyces cerevisiae* to take the lead. This high-concentration *Saccharomyces* batch is then added to the must of harvested grapes when ready. At this point, the winemaker is sending in huge populations of our desired yeast to the beginning of the main ferment, giving *Saccharomyces* a fighting chance to gain an overall competitive advantage much earlier in the process.

Spontaneous fermentations are slower and cooler than those made using commercial yeast, especially in the initial stages of fermentation. At the start, whilst wild yeasts are in play, not only does each strain have a part in building a different aroma and texture profile for the wine, but they also delay the start of an altogether quicker *Saccharomyces* fuelled fermentation. This enables more time for oxygen to react with phenolics in the wine, giving better colour stability and tannin polymerization. Cooler fermentations also lead to reduced loss of phenolics to heat – the more subtle volatile phenolic aromas are retained in the wine as opposed to being burned off in a warmer environment.

Nowadays we can also buy cultured strains of non-*Saccharomyces* yeasts, and even blends of yeasts for fermentation that can achieve better outcomes for winemakers such as lower acetic acid, higher fruit notes, more glycerol or higher lactic acid production. These commercial blends of yeast completely change the style of wine companies can produce, enabling winemakers to hone them into a more consistent wine character year on year.

Before we had the advances in technology to produce cultured strains of *Saccharomyces cerevisiae* commercially, wild yeasts were a part of the wine's characteristics and localized yeast demographics would play a huge role in altering wine styles. Wines differed greatly from one vineyard to the next and this was due more to local microbial communities than to the land itself. Could it be that we have lost this biological aspect of place by using commercial yeasts instead of indigenous ones?

Brettanomyces bruxellensis

'Brett', as the yeast is known to its friends (and foes), has become the talk of the town in trendy natural wine bars, though it was raved about in beer-making circles long before it became cool with the natural wine set. *Brettanomyces* is considered a spoilage yeast by most and it can add some largely disagreeable aromas to fermentation. In 1903 the Carlsberg Brewery found the yeast in an English ale, and went on to give it a UK patent, making it the first micro-organism in history to hold a patent. The name *Brettanomyces* aptly derives from the Greek for 'British yeast'.

Regarded as the black sheep of the wine yeast family, like its close cousin *Saccharomyces*, Brett also consumes sugar and converts it into ethanol. We regard the volatile compounds that *Saccharomyces* produces to be largely pleasant; Brett on the other hand metabolizes aromas more akin to animal, barnyard, and sweaty sticking plaster smells. Vomit and sewage are also descriptors accredited to Brett's handywork. The primary aromatic compounds *Brettanomyces* produces are 4-ethylphenol and 4-ethylguaiacol. The former is responsible for the sweaty sticking plaster smell, and the latter for a more pleasant spice and clove aroma. These aromas are utilized in a symbiotic relationship with fruit flies (*Drosophila*) since both the flies and their larvae are attracted to these scents and will assist in yeast dispersal, just as they do for *Saccharomyces*.

Brett can be present on grape skins and often finds itself in wineries and on equipment. It grows much slower than its speedy cousin and feeds quite happily on unfermentable sugars as well as the usual fructose and glucose; these sugars include those derived from oak and others that *Saccharomyces* is not able to consume. Like its cousin, Brett also has a high alcohol toxicity tolerance and because of this often rears its head at a much later stage in wine production, with higher risk in oak-aged wines. Brett's processing of these unfermentable sugars can lead to a thinner and less viscous style of wine, since these sugars give volume to unaffected wines. It is one of the few microbes resistant enough to grow on after *Saccharomyces* has come to dominate a ferment. A survivor through and through, it can even metabolize ethanol alcohol into acetic acid, making wine smell a little vinegary and changing the balance of esters. The two aroma compounds synonymous with Brett are formed by the synthesis of coumaric and ferulic acid, both of which are polyphenols found in grape skins and a big reason why Brett is more problematic in skin contact wines than white wines.

Many natural wine lovers will enjoy the additional complexity and sense of 'funk', earthiness, and savoury nature that Brett can add to the experience of wine, whilst others are repulsed by just the tiniest traces of it and will immediately declare the wine faulty. From my own experience of working with Bretty wines on the front line of service, I see its handiwork as an acquired taste. At the bars I've hired sommeliers with some impressive establishments on their résumé, many of which garnered Michelin stars and had award-winning wine lists. When training these polished soms on our wine lists the Bretty wines were the most polarizing. New recruits found it difficult to get their heads around these wines; they too saw them as faulty and were wholly unsure about most of what being branded a 'natural wine' entails. Months of training later, once tests are passed and they finally get into the full swing of service, they have the chance to talk about these wines with guests at every service and share their opinions openly on the Bretty wines with non-biased consumers. They take to retasting and with time slowly reform their initial views on the funkier styles of wine. I have seen first-hand how their opinions can completely flip. The same sommeliers proudly come in as guests on their nights off, always bringing wine-loving friends and family and actively choose the off-the-wall funky numbers to share and change initial prejudices within their group. They develop a taste for it, just like many of us do with beer and coffee as we age, many of us gaining an appreciation of these odd yet wonderfully counterintuitive tastes – in moderation of course.

Brett can be avoided at the winery by increased hygiene and thorough washing of grapes on the way in. High sulphur dioxide applications during, before and after fermentation whilst keeping the pH levels of the wine high will also keep Brett at bay. If a wine has been affected by Brett, it can be 'deactivated' by treatments such as filtration through dimethyl dicarbonate (DMDC for short – trading by the name Velcorin). DMDC is used widely as a disinfectant in the industrial beverage industry.

Lactic acid bacteria

Many thousands of years ago the role of fermentation in food was as a means to preserve whilst improving flavour. The most relied upon group of microbes to do this are a set of lactic acid bacteria that enable us to pickle vegetables, create yoghurt and make cheese. In east Asia they also play a role in the production of soy and fish sauces.

Lactic acid bacteria can also be utilized in wine production where their handiwork is referred to as malolactic conversion or malolactic fermentation. *Oenococcus oeni* is one form of lactic acid bacteria put to work converting the more tart malic acid into lactic acid and carbon dioxide. It's a process called decarboxylation, which enables the release of carbon dioxide. Because of this, some wines that may recently have undergone malolactic conversion may have a fizzy tingle to them.

Most distinctly, the flavour of the wine will be completely changed during this process. The green apple flavours of malic acid are replaced by sour, milky notes from the lactic acid. In addition to the lactic acid produced, most strains of lactic acid bacteria will also produce diacetyl from any citric acid present in the ferment. Diacetyl is a yellow liquid which results in flavours of cultured cream, butter and buttermilk. In turn, diacetyl can also go on to react with an amino acid in wine called cysteine to form thiazole, a light-yellow liquid which smells like hazelnut and popcorn. Both diacetyl and thiazole are pleasant in low doses, but in higher concentrations may not be so enjoyable and can be considered a fault. The production of diacetyl by lactic acid bacteria is related to the amount of oxygen, sugar and citric acid in the wine. It's a big reason why citric acid isn't used for the acidification of wines (adding acidity). Diacetyl formation in fermentation is encouraged by lower temperatures and the removal of active yeasts.

It was that crafty chemist Louis Pasteur who discovered the presence of lactic acid in fermentation. Local winemaker and beetroot enthusiast Monsieur Bigot asked Pasteur to investigate what was turning his wines sour. It turned out to be the opportunity Pasteur needed to make it big. In 1873 he published his memoirs on the topic saying, 'l'oxygène est le pire ennemi du vin', which translates as 'oxygen is the greatest enemy of wine'. He described how oxygen availability in fermentation inhibits ethanol production, driving yeast to switch towards aerobic activity; this became known as the Pasteur effect. This analysis uncovered the growth of lactic acid bacteria and other spoilage microbes which proved to be turning M. Bigot's wine sour. The discovery made Pasteur France's greatest scientist of the time and led to him meeting Emperor Napoleon Bonaparte in 1862. Soon after, he developed the method of pasteurization.

Lactic acid bacteria work best in warmer conditions (18–22°C), with a moderate pH of 3.3–3.5, and low doses of sulphur dioxide. Many winemakers will intentionally inoculate their wine with cultured lactic acid bacteria to ensure they get a strong start, enabling them to reduce the tart flavoured acidity of their wine and develop a rounder, creamier style. The bacteria found on the skins of grapes are less resistant to acidic conditions than yeasts would be, hence at crushing when the acidic grape juice is exposed to the skins, a lot of lactic acid bacteria will be lost. The start of fermentation as we know is highly competitive, and if sulphur dioxide is added, bacteria will be at a clear disadvantage, being less sulphur tolerant than yeasts. In most commercial winemaking practice this process of assisting *Saccharomyces*-led fermentation early on with sulphur dioxide is the norm.

Towards the end of fermentation, temperatures are naturally higher, acidity levels have decreased and free sulphur dioxide is largely depleted. This gives bacterial growth the upper hand. Here's where lactic acid bacteria will come into play, if there are enough survivors left over to repopulate or if the winemaker has inoculated with cultured strains. It is important for the winemaker to ensure all malolactic conversion happens before the wines have been bottled; if not, the wines will completely change from their desired style and become fizzy in the bottle due to the creation of carbon dioxide within a sealed environment. To prevent malolactic conversion, winemakers can cool the ferment, increase acidity or increase levels of sulphur dioxide. The addition of lysozyme enzymes will also limit the risk of conversion since they kill lactic acid bacteria. Red wines always go through malolactic conversion with a little colour loss as a result, whilst malolactic conversion for white wines is a style choice by the winemaker.

The process of malolactic conversion always increases the pH of the wine (making it less acidic) since lactic acid is a weaker acid than malic acid. It can be a brilliant technique for reducing the acidity of certain wines, if the flavour profile works stylistically.

Creamy whites and runny noses

Amongst the milky by-products of malolactic conversion, amines are produced in large numbers. These include both histamine and tyramine, which are formed by the decarboxylation of amino acids in the wine. We link histamines to allergies – a rise in histamines for allergy sufferers will lead to an increase in the sensitivity or severity of allergy symptoms. These include headaches, rashes, respiratory issues and light sensitivity – though it does depend on both the specific allergy and the individual in question. We are prescribed antihistamines when suffering from allergic reactions to block the effects of amines in our system, thus lessening the symptoms.

Histamines are also produced naturally in all fermented or aged food and drink. Parmesan cheese, cured meats and beer are all examples of histamine-containing edibles. It has become very popular to point to sulphur in relation to 'wine hangovers', but should allergy sufferers instead point to the histamines found in wine, especially those creamier whites and red wines which have all undergone malolactic conversion?

"Red wines always go through malolactic conversion ... malolactic conversion for white wines is a style choice."

Acetobacter bacteria

These anaerobic and airborne bacteria are everywhere in nature and found at their highest concentrations in sugar-rich environments such as nectar and damaged fruit. *Acetobacter* have a high tolerance for alcohol and do prosper in alcohol-rich environments because there are fewer microbial contenders around. In the presence of oxygen, they can break up ethanol into the sharp-tasting, vinegary acetic acid. Ethanol chemically shares one of its bonds with an oxygen atom. Acetic acid also has this one oxygen bond, plus an additional double bond with another oxygen. This explains why *Acetobacter* bacteria need access to plentiful oxygen to oxidize the ethanol into the pungent acetic acid.

If permitted, *Acetobacter* bacteria will have their way with the natural sugars found in wines and ciders and eventually we will be left with vinegar. Both *Saccharomyces* fermentation and malolactic conversion will also produce small amounts of acetic acid enzymatically but nowhere near the volumes *Acetobacter* can achieve, hence it is to be avoided in wine production.

There are a few other familiar micro-organisms that create acetic acid too. Wild strains of yeast enzymatically process a little more acetic acid than *Saccharomyces* would – unless *Saccharomyces* is stressed or nutrient deprived, which makes it produce more acetic acid than usual. Also, if our black sheep Brett is allowed to operate aerobically, it'll chip in and turn ethanol into a little acetic acid.

Since *Acetobacter* needs oxygen to grow, it has adapted to float, cunningly forming a thin biofilm on top of wine and endeavouring to turn it to vinegar from top to bottom at any opportunity. Following the advice of Monsieur Pasteur, we learn that oxygen is the greatest enemy of wine, so controlling the amount of oxygen available to wine is the most effective measure in curbing the growth of unwanted bacteria.

To note, the autolysis of yeasts supplies spoilage microbes with a good source of food, further enabling their development and growth. Yeast guts are full of nutritious amino acids, so wines like Champagne and many other autolytic styles run a higher risk of microbial instability. Regular applications of sulphur dioxide will kill *Acetobacter*, along with the other bacteria in the ferment. Dimethyl dicarbonate (DMDC) can also be employed as a chemical antimicrobial alongside sulphur dioxide under conditions of high bacterial risk.

Volatile acidity

One major ester produced by *Saccharomyces* is ethyl acetate. At low levels it can contribute to a familiar 'fruity' aroma in wine, though pesky *Acetobacter* produces much higher concentrations of these ethyl esters than we'd want. Higher concentrations are perceived as a wholly unenjoyable nail-varnish remover odour. When paired with the vinegary smell from acetic acid we refer to this as volatile acidity (VA), which is a fault in wine. Volatile acidity is technically measured by the concentration of low molecular weight, steam-distillable, fatty acids in wine, though by far the major contributor to the smell is the vinegary and nail-varnish remover aroma.

Competitive bioprotection

We've covered a lot of the good, the bad, and the workable microbes in wine production, and careful use of sulphur dioxide appears to be the sheriff in town when it comes to keeping the good microbes in charge, and the unwanted ones at bay. The sugary, nutrient rich, warm environment of fermentation creates a melting-pot of different microbial cultures and it's this diversity that leads to an array of volatile compounds, which may or may not be what we really want. One skilful method to get the best from our ferments without the use of sulphur dioxide is to get political, using the competitors to work against each other.

By using the right microbes, at the right time, we can deplete the resources from the ferment, reducing nutrients to lower the risk of microbial infestations further down the line. Two key elements of bioprotective practice include competitive exclusion and protective microbes.

Competitive exclusion

This is where a culture or blend of microbial cultures is selected to inoculate fermentation to outcompete other species or strains of micro-organisms. For example, the addition of selected strains of *Oenococcus oeni* at the end of alcoholic fermentation will ensure malolactic conversion takes place and prevents the build-up of Brett due to a lack of available resources. The addition of pure *Lactobacillus plantarum* strains into the grape must at the start of fermentation will help them gain a competitive advantage, protecting against more harmful moulds and the development of acetic acid bacteria.

Protective microbes

In this situation, microbes that produce compounds that can kill or deactivate other organisms are introduced to the ferment. These microbes may produce microbial killing agents such as alcohol or acids.

There's plenty that winemakers can do to bend the making of wine to their will, and as our understanding of microbiology and food science advances, so does our ability to manipulate the food and drink we make. In our next chapter we discuss the most common techniques in winemaking to harness the living world around us and create some order amid the chaos.

"Fermentation creates a melting-pot of different microbial cultures and this diversity leads to an array of volatile compounds."

6. Manipulations at the winery

Manipulations at the winery

Our ability to utilize the environment around us to our advantage has brought us precisely to where we are now. There's seemingly no end to our ingenuity and wine has been no exception. We're able to produce millions of identical wines, almost year on year for sale to supermarkets irrespective of vintage, climate or soil.

It makes the mind boggle to read through these last five chapters and fathom the complex interplay of factors needed to produce quality wine, and then to realize it can be overcome by technological intervention. These manipulations allow us to produce wine in the tens of millions of litres for distribution all over the world, for a somewhat indefinite shelf life. Just how do we do it?

Sulphur dioxide (SO_2)

We've been exploiting the antimicrobial effects of sulphur dioxide since the times of the ancient Greeks and Romans. Back then, it was commonplace to fumigate homes by burning sulphur to destroy the micro-organisms which cause disease and rot. Sulphur has been documented in winemaking since the 1700s and is now utilized at almost every stage of the conventional winemaking process, from harvest to crushing, and then fermentation to bottling. It's very handy stuff since it counteracts the two main causes of food spoilage by acting as an antimicrobial and an antioxidant.

There's been much debate over how exactly sulphur operates as such an efficient microbial assassin. Current studies settle on sulphur dioxide's ability to enter inside microbes and disrupt the activity of their enzymes and cell proteins. Only the molecular form of sulphur dioxide is small enough to enter through cell membranes, hence it is the concentration of the molecular sulphur dioxide that controls microbial growth. This microbial killer is also assisted by toxic alcohol in destroying microbes, therefore higher alcohol wines tend to require less SO_2 to keep them safe from microbial spoilage.

As an antioxidant, SO_2 works quite simply by reacting with oxygen before the phenolic compounds do, protecting the wine by sacrificing itself. It can also inhibit oxidative enzymes in a similar way to how it handles microbes, having the double effect of further protecting the wine from the enzymes which would cause oxidative issues.

When SO_2 is added to wine, it dissolves and reacts with the liquid. The short explanation for the bleaching effect we mentioned earlier is that it is caused by the SO_2 binding to anthocyanins, forming a colourless bisulphite. As more sulphur dioxide is added, more will dissolve into the wine and become what we call 'bound

SO_2'. Bound SO_2, once reacted with the wine, has very little of the antioxidative or antibacterial effects we desire: it's an inevitable loss that we cannot avoid. When the wine becomes saturated with bound SO_2 and can hold no more, a significant amount of useful 'free SO_2' is formed as more SO_2 is added to the mix. This is the molecular form we noted earlier.

The acidity of a wine has an exponentially dependent relationship to how much SO_2 gets bound, and therefore influences how much free SO_2 remains. Wines with high acidity are less sulphur soluble, so free sulphur dioxide is formed sooner, meaning less is required compared to those with higher pH levels.

Without any external addition of this preservative, a small amount of SO_2 is naturally produced as a by-product of fermentation anyway. Sulphur dioxide is classed as a toxin in food and is listed as one of the 14 allergens in EU law which needs to be present on food labelling. Anything over 10 milligrams per litre of sulphur dioxide must be clearly stated on the bottle; hence, the much talked about 'contains sulphites' declaration is found on almost every bottle of EU wine. Even without the artificial addition of sulphur dioxide in wine, levels just above the 10 milligrams per litre mark may be present in wine through the process of a fermentation alone. We often assume this label means sulphites are added to the wine, yet depending on the winemaking, technically this may not be the case.

"Sulphur is now utilized at almost every stage of the conventional winemaking process."

Are high sulphur wines the cause of a killer headache the next morning?

According to Allergy UK, less than 2 per cent of the population suffers from an allergy to sulphites, rising significantly to 5–13 per cent for asthmatics or those with respiratory illnesses. Sulphur often gets the blame for our hangovers and some people believe that organic wines may be more forgiving. From the data, this cannot be the case unless we are among the few individuals who suffer from a direct sulphur allergy.

Global laws regarding maximum sulphite levels in wine differ. Regions such as the EU, Australia and New Zealand set the maximum legal limit at 150 milligrams per litre for red wine, and 200 milligrams per litre for white wines. Sulphur dioxide is used widely in most food production, and in liberal amounts for dried fruit since a high sugar sliced product will be very susceptible to decay. In dried fruit we're talking about maximum permitted levels of 3,000 milligrams per kilogram within the same legal geographies, but to this day I haven't heard of sulphur getting the blame for a headache after a feast of dried apricots.

If we are to blame something for how we feel the morning after, it should be the combination of alcohol and histamines, with SO_2 coming in a distant third.

Surely 'natural' wines must mean no hangover?

The question I get asked most at aspen & meursault, and always with a glimmer of hope in the guest's eye that the morning after will be just that little bit more manageable. Knowing that the term has no legal standing, it means that 'natural wine' could mean a multitude of things. Not a great foundation to base a case study on. Let's assume for argument's sake that the natural wine in question has had limited intervention in farming and production, as the ethos implies. In the winery let's suppose something along the lines of biodynamic level restrictions on sulphur, just for this exercise.

In terms of the sulphur itself, we know that low or no added sulphur natural wines will only have a marked benefit on the minority of sulphur allergy sufferers.

What about amines then? Well, there's no conclusive evidence to show that histamine levels are any different in natural wines.

Next, let's put the chief culprit of the sore head on stage, glorious ethanol. It's the two-step process of breaking down the ethanol in our bodies that is the main cause of the symptoms we describe as a hangover. The first step takes place in the liver, when the enzyme alcohol dehydrogenase (ADH) gets to work at breaking down the toxic ethanol into acetaldehyde. Acetaldehyde is a little less toxic and can be the cause of dry mouth, headaches, increased skin temperature, nausea and many other symptoms associated with mild toxicity in the body. Acetaldehyde can also form in wine that has been around a bit too much oxygen, and in natural winemaking it is highly likely we're at higher risk of oxygen involvement if sulphur is out of the equation. Additionally, in a natural ferment, non-*Saccharomyces* yeasts such as *Kloeckera apiculata* can create more of this undesirable acetaldehyde. Personally, I am an advocate for low intervention wines, though counterintuitively the evidence suggests a higher likelihood of acetaldehyde in them, which when paired with ethanol won't feel great the next morning.

The second step of ethanol breakdown in the liver is the job of turning the less toxic acetaldehyde into an even safer acetate. The liver enzyme aldehyde dehydrogenase (ALDH) performs this task, and once converted to acetate, it will further break this compound down into water and carbon dioxide. Even though acetate is far less toxic, a lot of it will still cause a headache. The whole two-step process utilizes two lifesaving enzymes in making alcohol safe, but it does take time. It's why we dejectedly feel the repercussions the morning after. Now add an element of dehydration to the mix, and we can all remember how that feels.

The link between natural wine and hangovers is indeed a complicated one, since some factors may make hangovers worse, and others may make them less likely. Not the most helpful conclusion, yet as we are coming to learn, the science does not always leave us with the most gratifying conclusions.

Small side note: of late within the drinks industry there's been much talk about the 'discovery' of congeners as the cause of hangovers. This term is one created by the drinks industry as a blanket term for all the non-ethanol bits in alcohol. This includes acetaldehyde, tannins, histamines, acetate, all polyphenols, esters and more – it's a hefty list. In this same line of discourse, it's also often said that there are more congeners in darker alcoholic beverages, like red wine and dark rum, than lighter ones. This statement is of course true because of the additional polyphenols in the mix from either skin contact or barrel ageing. Using such a broad catch-all term to explain away a hangover seems a little bit of a con. It's so broad it can only be true!

What the hell does sulphur smell like?

Sulphur dioxide gas characteristically carries a pungent metallic odour, often described as flinty or likened to the aroma of a freshly struck match. If high levels are added during production, it can carry through to your bottle of wine and may be perceived by very sensitive tasters. These smells may not be wholly unpleasant and wouldn't be classified as a wine fault. Wine tasting notes of 'minerality' have of late proved very popular; whereas decades ago this was rarely mentioned, it now features prominently in the description of many a white wine. From a scientific standpoint, the origins of this commonly diagnosed mineral smell are elusive. Like all aromas, it is made more volatile by the presence of heat. Tests have shown that sulphur-based aromas in wine, such as struck match and flint, are more prevalent in wines that are served warmer.

When we discussed soil in earlier chapters, we concluded that most rocks don't smell, so what is it about struck flint that gives us that sulphuric smell? How are sulphur and flint connected? Flint is derived from chalk in a complex process that began in the chalk seas millions of years ago. We could spend a whole chapter on this, but to mitigate the risk of losing you we'll speed up the explanation. Under very specific acidic conditions silica replaces the calcium carbonate in chalk, grain by grain. Local underwater bacteria in the chalk sediment operate anaerobically and produce hydrogen sulphide that meets dissolved oxygen to form an acid and create flint. It's this hydrogen sulphide by-product in the flint from the ancient microbes which, when stuck with force, will ignite and oxidize to create these low-level 'flinty' smells.

This is not to say that the descriptor 'minerality' comes solely from the addition of sulphur dioxide. We learned that sulphur is also produced naturally by yeast in fermentation. The term for the volatile sulphur compounds detected in wine is reduction. We will go deeper into the nuances of reduction later in this chapter but for now it's helpful to think of it occurring most in low-oxygen environments. In reductive conditions like completely sealed bottling or oxygen starved winemaking, our yeasts are prone to producing more sulphuric compounds like hydrogen sulphide and thiols. These can present as pleasant mineral, flinty, struck-match aromas.

If the balance tips, higher concentrations of volatile sulphur compounds could result in wholly unpleasant rotten egg, garlic or cabbage-like aromas. These are smells most would associate with sulphur outside of a nuanced wine context and if discovered in wine are undoubtedly a fault. It's the hydrogen sulphide that contributes most to the rotten egg smells. From an evolutionary standpoint we're all finely tuned to picking up on these types of smells at extremely low odour thresholds – less than one part per billion! Those of our ancestors who could better detect the sulphurous smell of rotten food would inevitably go on to live longer and prosper.

Fiddling with fermentation

There's a lot that can be done with a ferment to change the outcome of our wine. Here are a few of the cunning manipulations which help us achieve the wine styles we desire.

Grape must adjustments

Before we start fermenting, adulterations can be made to the must which will sway the style of our wines. These include enrichment, adding tannins and acidification.

Enrichment

Enriching the must refers to the addition of sugars to the must, ordinarily to increase the final alcohol by volume (abv) of a wine. This is commonplace in less sunny and cooler climates where vines can struggle to photosynthesize enough to generate optimum sugar ripeness. Sugar is often added as dry sugar, grape must concentrate, or rectified concentrated grape must (RCGM) – a flavourless industrially processed grape syrup. The practice of enrichment has been going on for quite a while now: in 1662 English scientist Dr Merret noted how Champagnes could be made more sparkling by the addition of sugar before fermentation.

The technique of adding dry sugars to must can also be called chaptalization, after the French chemist Jean-Antoine Chaptal. He began his career running factories which produced industrial quantities of sulphuric and hydrochloric acid during the French Revolution. At the time, the wines of northern France tended to produce high acidity, low alcohol wines due to the colder climate and lower sunlight hours compared to Chaptal's hometown in the south-west. For centuries many winemakers in northern France had boiled down the grape must to 'cook off' the acidity and concentrate the must to achieve higher alcohol wines. In post-revolution France, the fuel required to do this was scarce and pricey. Chaptal initially came up with the idea of adding plaster of Paris to the must to reduce the water content, before hitting upon the far more sensible process of chaptalization, extracting sugars from beet and adding this high sucrose and low-cost dry concentrate to the must.

In the EU this practice is only permitted within certain cooler climate regions. Some hotter climates are also allowed to utilize RCGM within select PDOs. Wine must can also be concentrated by reverse osmosis or cryoextraction – which involves freezing the wine and removing the ice particles. If winemakers wish to lower the alcohol of their final wines, some regions allow them to add water and dilute the must prior to fermentation. This lowers alcohol levels, but also dilutes the other desired characteristics of a wine.

Adding tannins

Tannins can be added to the must before fermentation or to the wine afterwards to aid in clarification, enhance mouthfeel or to colour stabilize skin contact wines. Tannins are often added in a powdered form, industrially processed from grape skins or other natural sources like tree bark.

Acidification

In hotter climates, grape malic acid levels can drop quite significantly. Acidification is commonplace in warm climates, where acids are added to wine or the must to top up acidity levels. It's most often that DNA-helix-inspiring tartaric acid (see p. 106) which is added in a dry powdered form. Citric and malic acid can also be used but, as we learned earlier, they change the taste of a wine and promote lactic bacteria, so aren't ideal. High acidity environments can denature certain microbes, hence high acidity wines are more stable for long term ageing and general shelf stability requirements. More lactic acid may also be added after a wine has gone through malolactic conversion to promote a softer style of acidity. In the EU, acidification is only permitted in hotter regions, generally those in the south of Europe.

Deacidification can be performed by the addition of calcium carbonate (chalk) or potassium carbonate to must or wine. There are also stipulations within EU law regarding the extent to which this practice can be employed.

Skin maceration

Time on skins is fundamental to creating the type of red, orange or rosé style we're after. The longer on skins, the more colour, tannins and good bits from the exocarp we extract into our wine.

Cold macerations occur before the fermentation, in a high sugar, zero alcohol environment. The must is chilled to reduce the rate of oxidation and bacterial spoilage, and to prevent fermentation from kicking off before the winemaker wants it to. Usually this can occur for up to a week prior to fermentation and is a gentle way to specifically extract colour from the skins, since anthocyanins are highly water soluble. In the nineteenth century the use of lamb's blood to dye wines was widespread. It's no longer a permitted wine practice so for now we'll have to rely on our grape skins for colour.

Post-fermentation macerations occur in a high-alcohol, low-sugar environment. The winemaker can still extract colour in these conditions due to the continued presence of water in the wine (since anthocyanins are water soluble), yet we also see a significantly higher rate of tannin extraction from the skins since tannins are very soluble in alcohol. Post-fermentation macerations need to be monitored tightly to ensure tannins aren't over-extracted.

"The longer on skins, the more colour, tannins and good bits from the exocarp we extract into our wine."

Thermovinification and flash détente

These are industrial methods for speeding up extraction from the skins by applying temperatures of up to 95°C to the grapes. With flash détente, the grapes are then cooled rapidly in a vacuum after heating. The whole process takes less than two minutes but is so intensive it causes the individual cells in the grape skins to burst, making it a super-fast extraction procedure. High-tech indeed, however it risks significant loss of aroma due to the levels of heat involved and can result in cooked aromas in the wine. The equipment is expensive so is mostly used for the production of bulk wine, where time and quantity are of the essence. The silver lining for bulk wine producers using these processes is that the temperatures are so high they kill bacteria and botrytis, and the heat denatures the laccase enzymes that cause oxidation. When you're mass producing wine it's hard to account for the health of the sheer volume of fruit, so this method is a real bonus, ensuring extraction, speed and stability for bulk wine producers.

Skins float, so for all maceration methods there's an element of physically finding a way to get the skins back in the liquid. If they're left to float at the top of the tank, we don't get to extract all the good bits we want from them and we risk nutrient-rich skins harbouring spoilage bacteria like *Acetobacter*. There's a multitude of ways to keep the skins down, and the more movement there is, the faster the extraction.

Why are orange wines so trendy now?

We sell as much orange wine as we do white wine at the bars. Guests are either ardent orange wine devotees or curious white wine drinkers who want to give them a chance. In the early 1990s this wouldn't have happened, so is this hip style of white grape skin-contact wines a new invention? Put simply, it's quite the reverse.

Orange wines were some of the first wines ever made, produced at a time when we didn't have the technology, know-how or desire to separate the skins from the ferment. In an era without refrigeration, flavonoids in the skins had a very handy antioxidant effect, which would have extended the shelf life of 'white' wines without the use of sulphur. In medieval Europe these wines were also very much in vogue. We can see this in many older Renaissance paintings, which show orange and not white wines in the glasses of merrymakers.

Whole cluster fermentation

Whole cluster fermentation describes an environment where the grapes are kept whole and on stems. It's a winemaking tradition used extensively in Burgundy with their Pinot Noir and on occasion in the northern Rhône with Syrah. With this style of fermentation, we're including stems and whole grapes, including the seeds! This provides large quantities of tannins to the mix, and for wines like Pinot Noir, which lack acylated tannins, the added tannins do well to keep better colour stability.

Carbonic maceration

It does feel like we're getting acquainted with our yeasts quite intimately, understanding how they work through the good times and the bad, but for the first part of carbonic maceration let's put our yeasts to one side, just for a moment. There is an alcohol-producing phenomenon which occurs without the need for *Saccharomyces*.

For this enzymatic reaction, we require our grapes whole, just like out in the vineyard, where the sugars inside the grapes are not exposed to external yeasts. In this whole grape state, nothing happens, it's just grapes. The fun begins when we create an anaerobic environment and take all the oxygen away. This is performed at the winery by filling a tank of uncrushed grapes with carbon dioxide, displacing all the oxygen in the tank.

Now that we've removed oxygen from the scene, weird things start to happen, and certain enzymatic machinery comes to life. These enzymatic reactions occur within the cells of the grape and produce alcohol, alongside some other very distinct compounds. As alcohol develops during this anaerobic enzymatic intracellular activity, it reaches a point where the alcohol volumes are too large for the cells to contain, and they explode! This enzymatic process creates some incredibly unique flavours, different from those created by traditional yeast fermentation. These include candied fruit, kirsch, musky, cinnamon or oak-like aromas. Regions such as Beaujolais are renowned for their carbonic and semi-carbonic styles of wine that carry these characteristics. Carbonic wines are reported to mature faster, and create softer tannins.

"This enzymatic process creates some unique flavours, different from those created by traditional yeast fermentation."

So what exactly is happening here? With full carbonic maceration, where carbon dioxide is used to fill the tank, the grapes absorb the CO_2 and fill themselves with up to 50 per cent of the stuff. The enzymes within the cells shift from aerobic respirational activity to anaerobic activity, where an enzyme called malic dehydrogenase kick-starts the metabolism of malic acid into ethanol, succinic acid and aminobutyric acid. With malic acid being consumed in this process, we're left with the two other weaker acids, citric and tartaric, an overall net decrease in acidity. The ethanol produced by malic dehydrogenase will reach around 0.5–2.2% abv and the cell membrane's integrity will eventually give way and burst open.

During this enzymatic activity, high levels of glycerol are produced, as one can imagine from what is altogether quite a stressful environment. Shikimic acid is also created in the process and promptly degrades into benzaldehyde (presenting flavours of cherry and almond), vinylbenzene (styrene and plastic smells) and ethyl cinnamate (cinnamon, strawberry and honey aromas). It's this aromatic blend which gives carbonic wines their truly distinctive flavour characteristics and voluminous nature.

When fermenting carbonically, exocarp phenolics like anthocyanins and tannins are lowered due to the reduction in skin surface area exposed to the must for the duration of fermentation whilst the grapes are whole, compared with when they are crushed. An added cause of low tannin content is the lower levels of alcohol during the start of fermentation which do not come in contact with the skins, delaying the start of tannin extraction. Some winemakers choose to perform whole bunch carbonic maceration as the stems do a good job at supplementing the tannins in their wine.

Semi-carbonic maceration utilizes the same anaerobic enzymatic activity, but instead of excluding oxygen by use of carbon dioxide, we simply submerge the whole grapes under a layer of grape must, where oxygen levels will drop below 5 per cent. In this scenario, there's a portion of must fermenting with yeast in the grape must, and a portion of whole grapes fermenting enzymatically from inside, all at the same time!

When the cells of the carbonic grapes under the grape must hit approximately the 2% abv mark they explode, and more liquid is instantly added to the mix. The must level in the tank increases, and a higher tier of whole grapes submerges deeper into the ferment. This chain reaction continues layer by layer until the entire fermentation is finished off by *Saccharomyces*.

Utilizing the lees

Lees is the fancy winemaking term for the sediment left at the bottom of fermentation, an assortment of dead yeast, bacteria, precipitated tannins, bits of grapes and other solid compounds that aren't able to dissolve into the wine. Bigger parts of this assortment are formed quite quickly, materializing in just a day of fermentation and are referred to as *gross lees*. The smaller fine particles take longer to settle and are aptly dubbed *fine lees*.

We touched on the self-destructive effects of yeast autolysis and the bready, yeasty aromas they can confer whilst adding texture to the wine. These attributes, if desired, can be amplified by increasing the amount of contact these autolytic yeast cells have with the wine. This can be done in many ways, which include stirring up the yeast with a big paddle, referred to as *bâtonnage,* spinning the barrel to unsettle the resting lees, or in the case of Champagne employing a 'riddler' to hand turn up to 40,000 bottles a day in order to rearrange the dead yeast pyramids that settle inside the bottle. In most cases the process requires additional manual labour and may factor in as an added cost to the wine. All sparkling wines made by the 'traditional method' will rely on lots of lees contact, hence you can expect these wines to taste exceptionally bready, toasty and yeasty.

Increasing yeast autolysis has the added effect of softening tannins in wine as the autolytic debris can polymerize with the tannins, forming longer chains and smoothening their mouthfeel. This debris also provides nutrients for other micro-organisms and can assist the growth of unwanted lactic acid bacteria or *Brettanomyces* in one's wine, so winemakers need to be careful. Although there's nothing that a bit of good old SO_2 can't fix.

Utilizing wood

Wooden barrels are so often what comes to mind when we think of making wine: it's certainly a more romantic image than stainless steel tanks or concrete eggs. Wood ageing contributes largely to the character of a wine, and different sorts of wood will do so in differing ways. Oak is most often used, however acacia, cherry, and even chestnut can also be utilized in wine ageing or fermentation.

All types of wood ageing will change a wine through slow oxidation, where the pores in wood allow micro-oxidation to occur, something that wouldn't happen when using stainless steel. Small amounts of oxygen intake can be good for wine, if controlled. Micro-oxidation intensifies the colour and the stability of red wines by catalysing reactions between anthocyanins and tannins, and increases the rate of tannin polymerization, lengthening the DP and softening the mouthfeel of tannins.

Barrels are often said to breathe, and in a way it's true. Not only do they take in small amounts of oxygen from their surroundings, but wood barrels also serve as a medium for the outward transfer of carbon dioxide, ethanol and water evaporates through the joints in the barrel staves and head. The speed of exit for this evaporation is dependent on environmental factors and the age of the barrel; newer wood will absorb wine at a faster rate. The evaporating wine escaping from the barrels causes the level of liquid within the barrel to drop and the air gap at the top of the barrel to widen. This can cause further negative oxidative effects to occur if not topped up with fresh wine. Most oxygen ingress is from around the bung of the barrel, and the positioning of the bung on the barrel plays a big part – top barrel bungs allow less oxygen interaction with the wine than side bungs!

"Wood ageing contributes largely to the character of a wine, and different sorts of wood will do so in differing ways."

Breaking down wood and wine

A typical woody biomass is largely made up of 40–50% cellulose, 25–35% hemicellulose, 20–40% lignin and 5–10% tannins.

Cellulose, which forms much of wood's structure, does not alter much in the winemaking or barrel-making process. Softwoods often contain a lower proportion of hemicellulose and a higher proportion of lignin than hardwoods, which influences the style of flavours we derive from the wood. Within the structure of wood, we also have fatty and oily lipids, abundant in the cell membranes of trees. When wood is seasoned, these lipids fall apart to form lactones; different types of wood will contain different sorts of lactones. Oak, for example, carries sis- and trans-lactones which are perceived as coconut, celery or raw wood aromas and form part of the flavour of oak aged wines.

Hemicellulose is a gluing agent within the wood, made up of sugar molecules such as arabinose and xylose. When they come into contact with wine these molecules can hydrolyse and break down into smaller pieces. If heated to over 200°C (as in barrel making), hemicellulose will decompose into even smaller constituents like creamy diacetyl, furfural, sotalon, furanones and maltol. The last four constituents mentioned contribute to caramel, malty, spicy and nutty smells which will be conveyed to a wine on barrel ageing. These oak heat treatments are achieved by seasoning and toasting the barrels to disinfect the wood and break down hemicellulose and lignin before wine ageing.

Lignin is a stable hexagonal carbon structure that requires a bit more energy to break down. It will start to pyrolyse at around 300°C, at which point it will fall apart into compounds like vanillin, guaiacol and eugenol – responsible for smells of vanilla, roast coffee and clove respectively. These aromas are utilized in nature by clove trees and vanilla plants as chemical and antimicrobial defences. Fun fact: vanillin was the first flavour molecule to be manufactured industrially!

Tannins found in wood and plant stems are slightly different from those found in grapes. We refer to them as hydrolysable tannins as they're easily hydrolysed by heat or acidity, meaning their structures can be cleaved easily into smaller constituents, in this case gallic or ellagic acids. These acids go on to form tannic structures like gallotannins and ellagitannins, which will end up in your wine. Gallotannins have a glucose molecule as their main structure and form shorter DPs which present as more astringent. Ellagitannins are widespread in nature, are highly reactive and love grabbing onto other flavonoids, forming longer, smoother chains. They do very well to stabilize colour by binding to anthocyanins.

The top three oak species we use in wine are pedunculate oak, sessile oak and American oak.

Pedunculate oak or *Quercus robur*

These trees originated in French forests and are used extensively in Limousin, Burgundy and other parts of southern France. The pores in all wood are formed by the xylem, long, hollow transport links for water and nutrients through the tree. If sawn, these pores will allow liquid to leak through the barrels, rendering them useless, hence this type of oak needs to be split instead of sawn. Much skill is required to form splits along the direction of the xylem rather than against it, to create a watertight and relatively pore-free structure. The lack of surface pores reduces the relative surface area of the French oak barrels, thus lessening the barrel's effect on the wine. The craftmanship required in splitting oak is far more expensive than sawing, which is a huge factor in why French oak is far pricier than American oak. Genetically, pedunculate oak contains very high levels of ellagitannins but carries low levels of extractable aromas, meaning it lends itself well to the long-term ageing of wine.

Sessile oak or *Quercus sessilis*

This species also originated in French forests and is often used in central and southern France. Sessile oak grows more slowly than pedunculate oak. Analyses show that sessile oak contains more aroma, in particular lactones and eugenol, but lower levels of ellagitannins than pedunculate oak. The easiest way to differentiate these two French oaks is by the leaves: sessile oak leaves have long, thin stems at the base but pedunculate oak leaves have little 'ears' at the base. Sessile oak also requires splitting, so also makes for a pricier barrel.

American oak or *Quercus alba*

This fast-growing American species most notably contains tyloses within its xylem structures. Tyloses are little overgrowths that form as nodules inside the xylem tubes during conditions of stress or pathogen attack; they're designed to slow down the spread of dangerous pathogens in the plant. These nodules partially block the xylems and slow down the flow of liquid. When it comes to making oak barrels, tyloses mean the wood can be sawn, as they block seepage through the pores. This reduces overall wastage in the making of the wood barrels and makes the whole production quicker and cheaper. The sawn American oak surfaces are cut to make the most out of the tree, regardless of the direction of xylems. The surface of the barrel will therefore have lots of pores available for the liquid to fill. This effectively increases the surface area of the oak, intensifying the aromatic oak effects on the wine. American oak is genetically predisposed to containing higher levels of lactones than French oaks so will contribute more coconutty and woody aromas to our wine.

There are also many other barrel parameters which serve to increase or decrease their effects on wine. These include the level of seasoning and toasting of the barrels used, the number of times the barrel has been used, the size of the barrel and, of course, the time spent in barrel. Many new winemaking regions have adopted the use of staves and wood chips which are significantly cheaper than an oak barrel. Staves and chips significantly increase the surface area of the oak on wine, whilst keeping costs down. If too heavy handed with oak a winemaker may ruin the wine or worse, tarnish the winemaking reputation of a region. This was the case with California, where only of late are they shaking off the bad reputation of making heavily oaked wines.

Clearing things up

Cloudy looking wines have become synonymous with the natural wine scene. It's not a trend unique to wine: craft beers and ciders have been doing this for some time now, and this is what the first iteration of wines, beers and ciders looked like before we had the technology or care to bother removing the cloudy bits for the sake of aesthetics. Most wines we buy now are clear and contain few or no hazy bits. The real fear in leaving our wines cloudy is that over time we simply don't know how these bits will react with the wine. They may cause wines to go bad or cause bottle-by-bottle flavour variations, which over time will only drift further apart. For mass-produced supermarket wines it's certainly not what we would expect, though for a smaller batch boutique wine we may not be so fussed. In the case of many natural wines and pét-nats (a low intervention sparkling wine method) we don't plan to keep them on the shelves for too long – the quicker we drink them, the less chance of variation occurring over time. All living products will also vary over time and natural wines are seen as living products which aren't meant to be as time hardy as conventional wines.

The two steps in making our wines completely clear and debris free are collectively known as *clarification* and *fining*. Clarification can be performed before or after a wine is fermented, in several ways:

Sedimentation

The oldest and simplest method of making our wines a little clearer: we wait. With time, the undesirable bits in the wine will fall to the bottom of the tank, then we just tap off the wine at the top. The longer we wait, the clearer the wine will be. This technique can also be referred to as settling.

Clarifying additives

Compounds can be added to the must to speed up the process of sedimentation. Pectolytic enzymes are usually added to break down pectins in the grape must. Pectins are the structural acids found in plant cell walls; they're used in cooking as a thickener. Jam-making relies heavily on them to turn fruity liquid into a congealed jam state. By introducing pectolytic enzymes to break the pectins down in the must, we're speeding up the separation between the liquids and solids thus aiding sedimentation.

Flotation

Here we do the opposite of waiting for the dust to settle. Gas is bubbled through the bottom of our must, forcing particles upwards. Winemakers will then add a binding agent so the bits can easily be skimmed off the top. If oxygen is used as the gas, this can also be a clever way of hyperoxidizing the must, a technique mostly used for the clarification of white wine.

Centrifuge

Shoving the must or the wine in a giant spinning centrifuge will rapidly separate the solids from the liquid!

Filters

Filters of varying materials and coarseness can be used to siphon off some of the larger bits in wine or must. These range from coarse cloth filters to sterile filters that are made of stainless steel and can trap the tiniest of bacteria and yeasts.

"Natural wines are seen as living products which aren't meant to be as time hardy as conventional wines."

Hyperoxidation

Though it seems counterintuitive to shove large amounts of oxygen through the precious grape must, this forced oxidization serves to reduce bitter and astringent phenolics in the final wine by reacting them with oxygen. Hyperoxidation is a means of oxygen-hardening the must, making the final wine less likely to react with oxygen in future. This has a sizeable effect on the browning potential of a wine too. Just like a freshly cut apple, when wine reacts with oxygen, it will brown. With hyperoxidation, we force the browning enzymes in the must (like tyrosinase) to do all their browning beforehand. The hyperoxidized juice becomes very dark brown in colour, then we siphon off the brown bits at the top and we're left with a very lightly coloured white wine which is less likely to brown and oxidize later.

After clarification

Even after clearing up the wine there may still be incredibly small hazy particles in the mix: a process called fining is often needed to remove them. As the name suggests, this is about taking the finer particles out of the liquid and is almost always performed after fermentation. This step of winemaking will also define whether our wine is vegetarian or vegan, or not. There are many ways to fine a wine, and all involve adding something that is heavy enough to sink, attracts those tiny particles and imparts no flavour. Below are some widely used fining agents.

Bentonite

We came across this clay earlier. In fining it's sprayed over the top of the wine, and as it sinks it'll adsorb tiny particles and unstable proteins. It has little influence on the flavour or texture of the final wine and is a vegan friendly way to get rid of the haze.

Vegetable proteins

Usually from potatoes, beans, soya beans or chickpeas. Another vegan option for fining a wine.

Egg white

Back in the day (we're talking eighteenth century), this was the most used method of fining. It can be added as dried powder or in its natural gloopy form, where it sticks to tiny particles as it sinks to the bottom. Since it's a protein-based fining agent, it will also bind with tannins, increasing DPs and making the wine feel smoother and less astringent. Technically our wine, while vegetarian, is no longer vegan, however there is a train of thought that says that because none of it is left in the final wine product, the wine itself is still vegan. It's an ethical question and very much depends on where you as an individual draw the line.

Isinglass

This substance is obtained from the dried swim bladders of fish (yes, fish innards). It's mainly taken from our most favourite of egg-laying fish, the mighty sturgeon. Isinglass is a form of collagen and works in a very similar fashion to egg white. It's quite a bit cheaper than eggs and has therefore become the more popular industrial fining agent in drinks.

Gelatine

Another collagen, this time from pork. Who would have thought it? It sounds rather extreme, however it was only in 1997 that the EU and USA banned the use of ox blood as a fining agent in wine, and only then as a reaction to the mad cow disease scare.

Keeping things stable

Most of our manipulations in wine are endeavours to keep it all stable and consistent for the long term. In a global, large-scale economy, food and drink stability are requisite, and wine is no exception. Lucky for us, technology has stepped in to ensure we can deliver. There are many ways in which a wine may appear inconsistent or unstable: two we need to monitor closely are microbiological stability and tartrate stability.

Microbiological stability

In bulk wine production having tiny living microbes in our wine could lead to disastrous outcomes. Most microbes struggle to live in wine because of the inherent high acidity and alcohol content, and the more of each we have in wine, the more microbe proof the wine will be. There are still a few alcohol and acid tolerant yeasts like *Brettanomyces* which can rear their ugly heads after bottling, even bacterium like lactic acid bacteria and *Acetobacter* bacteria will have a go if you give them the opportunity. Sugar-rich environments promote microbial growth, so wines with residual sugar and sweet wines will need some extra protection in place. The usual protective go-to, you guessed it, more sulphur dioxide! Sorbic acid can also be added to prevent yeasts from growing. Sterile filtration is an effective method to remove yeasts and bacteria entirely, and often the addition of dimethyl dicarbonate (DMDC) can be used to deactivate *Brettanomyces*.

Tartrate stability

As you will recall, tartrates are those little wine crystals that sparked the discovery of the DNA double helix, and some people find them offensive. Many guests at the bars have stood up, walked across the room eagerly with crystals in hand that by the time they've reached me have all but melted away in the heat of the moment. They're a natural part of wine and won't affect its flavour.

The crystalline tartrates which form are predominantly potassium bitartrate and sometimes calcium tartrate. When a wine is chilled, they become less soluble and will crystallize around any available potassium or calcium landing pads, only to swiftly melt away again when popped on a warm finger. Global wine consumers aren't big fans of tartrates so producers have had to find ways to get rid of them before bottling the wine for loading onto shop shelves.

A common way to remove potassium bitartrate from the wine is to chill it, let the crystals form, and then filter them out. Finer particles will need to be removed before this step as tiny particles, such as natural proteins, may prevent all of the crystals from forming. To speed up this process of cold stabilization, additional potassium bitartrate can be added to the wine to catalyse the crystallization part, before it's all filtered out together. Other industrial methods include ion exchange, which swaps potassium and calcium ions in the wine with hydrogen or sodium ions to stop tartrates forming entirely, or by adding carboxymethyl cellulose from wood, which prevents tartrates developing to a perceivable size. Another cheeky industrial method often used adds metatartaric acid to the wine, which stops tartrate crystals growing, but only for a short period of time.

We have indeed become experienced manipulators in wine. I am certain that readers are fully aware of and are completely fine with much of what we have mentioned, but some of the industrial chemical additions and techniques, utilized for the most part in mass wine production, may have surprised you. It's easy to turn our noses up at the use of these methods, and at bulk wine, but we should remind ourselves of why they have come to exist in the first place and what we're trying to avoid – whilst keeping costs down of course.

"In a global, large-scale economy, food and drink stability are requisite, and wine is no exception."

Faulty wines

We conclude this chapter with some of the wine challenges for which remedies still do not exist: our technological advances have not yet permitted us to overcome them, either for cost reasons or due to moral and health concerns.

Oxidation and volatile acidity

We know the effects of micro-oxidation in quite some detail now, and it turns out a bit of oxygen is a good thing. During fermentation the presence of oxygen is a necessity if we want to avoid carbonic style characteristics, and small amounts are also needed for *Saccharomyces* to work at their best. On bottling, a tiny amount of oxygen ingress will also do the wine some good, plus the slow release of oxygen a cork closure provides will catalyse the wine to evolve with time. The effect of nano-oxidation through cork is highly valued; these changes would not occur when using sealed enclosures like plastic corks, wax sealed tops or screw caps. The boutique array of aged underwater wines would also limit the amount of oxygen ingress into the wine. We sell some of these unique wines at the bar, they're gorgeously decorated with the remnants of sea life and drink exceptionally. The reduced oxygen ingress, constant sea movement, and increase in pressure as the cork is sucked back into the wine at the bottom of the sea bed all make a difference to how the wine will age.

But when does micro-oxidation become full on oxidation? It's a fine line: if too much oxygen is allowed into the wine at fermentation or after bottling, we risk *Acetobacter* bacteria getting a hold and producing acetic acid. Their vinegary aroma can add a subtle complexity in small concentrations, but if it gets out of hand we're left with a vinegary wine. Damaged fruit coming in at harvest can also be the cause of volatile acidity in a wine, where fruit is infected pre-fermentation; in bulk production it's harder to manage the quality of fruit and more sulphur dioxide is often the solution.

The situation caused by spoilage bacteria and the vinegary acetic acid can be worsened by the esterification of the ethanol in our wine, which combines with the acetic acid to form the ester ethyl acetate. We learned how yeasts create fruity esters, which are formed from free alcohol and acids, but in this case we've produced a pretty nasty ester that smells like nail-polish remover. Ethyl acetate is used in industry as a solvent for paint strippers, glues and nail-polish removers, and this is precisely what it smells like in our glass. It is best avoided entirely by preventing the build-up of acetic acid bacteria in the first place.

When oxidized wines change colour, they shift to a shade of brown. Chemical oxidation will mainly occur in wine without the participation of enzymes, where any oxygen present will oxidize wine polyphenols, turning them into quinones that then form insoluble brown compounds. In this process, the oxygen is reduced to hydrogen peroxide (H_2O_2) which reacts with the ethanol to form acetaldehyde, the compound responsible for the bruised apple or nutty smell. In the worst cases, oxidized wines will give off odours of all three components: nail-polish remover, nutty bruised apple and vinegar.

Reduction

In scientific lingo reduction is the opposite of oxidation, but in wine it means an entirely different thing. It refers to the volatile sulphur compounds in the wine, things like hydrogen sulphide which create horrendous eggy and sewage-like smells. We also mentioned how, in smaller concentrations, these compounds may give pleasant flinty or 'mineral' aromas.

Reduction is largely caused by yeasts working under stressful conditions, and most often due to a deficiency of a nutrient such as nitrogen or oxygen. Chemicals like hydrogen sulphide can also come about from the decomposition of larger solids in the ferment like the gross lees. Removing the gross lees early on can prevent this from occurring. It's a little confusing when using the word reduction to describe this wine fault, not just because it has two meanings, but because there is also some correlation between reductive characteristics (eggy, etc) in wine, and if the wine is made reductively (with very little oxygen present). Recent investigations have discovered that the cause of reductive, eggy characteristics is down to the particular strain of yeast utilized in fermentation and their stress levels due to a lack of oxygen or nitrogen, as opposed to the wine itself being reduced in any way – in the scientific sense of the word. I told you this was confusing.

Cork taint

Cork taint is a common fault, so common, in fact, that it often takes the blame for other faults. Corked wines are caused by cork closures infected with something called 2,4,6-trichloroanisole (TCA) that gives our wine a musty, damp cardboard or muted aroma.

TCA is formed in the bark of the cork tree. Within all tree bark, small pores exist, and within these pores many colonies of bacteria and fungi live. The fun-loving bacteria like penicillin and aspergillus which make up these colonies come up against some pretty brutal life challenges, and just one of these happens to be a set of newly developed pesticides and fungicides (based on the synthetic active ingredients chlorophenol and halophenol) which were rolled out between the 1950s and the 1980s. These now-banned 'forever' chemicals have broken down so slowly in the soils that they're still present to this day. When in contact with these dodgy chemicals the bark bacteria switch on a defence mechanism which renders the fungicides harmless but creates TCA as a by-product.

Another way TCA can come about is via 2,4,6-trichlorophenol (TCP) which is generated when wood lignin encounters bleach (sodium hypochlorite). Fungi and bacteria will then convert TCP into TCA.

We humans have an incredible ability to detect the teeniest amounts of TCA, sensing it at levels as low as two parts per trillion. That's like smelling a teaspoon of oil in the equivalent of a thousand Olympic sized swimming pools! What's fascinating about TCA is not only the distinctive, musty, damp-cardboard smell, but also its ability to suppress our olfactory signalling and physically limit our ability to smell. This is why we believe corked wines to have lost all of their vibrant fruity related smells – it's not the fruit aromas in the wine which have gone, but our ability to smell them. Aside from affecting our enjoyment of wine, the effects of cork taint are harmless.

The development of the screwcap closure was spurred on by the director of Australia's Yalumba winery in 1964 as a reaction to all the Australian wine that had to be chucked because of dodgy European corks. Portugal has always led the cork-supply business, and a fair amount of infected Portuguese cork ended up on the other side of the world. Screwcaps only really caught on in the 1990s, when cork taint really became prevalent in wine, due to the widespread use of those chlorophenol and halophenol based biocides.

Smoke taint

This fault is caused by the presence of smoke during veraison, just when the grape berries start to change colour. It's a huge issue in areas predisposed to forest fires as wines affected by smoke taint develop undesirably ashy, smoky and burned flavours. The main compounds responsible are ones we've already found from oaking: guaiacol and methylguaiacol. This time the compounds develop in the skins of the grapes; hence smoke taint is mainly an issue in skin-contact wines. Teinturier varieties are a great way of skirting these issues due to their dark flesh and ability to make red wines without the need for skins.

Smoke taint is caused by the grape's defence against toxic smoke compounds. Grapes protect their insides from toxifying by rapidly binding the smoke toxins with sugar molecules in the skins, a process called glycosylation. Once affected it can't be undone. Common smoke-taint affected regions include Australia, South Africa and, increasingly, California and Oregon.

Eucalyptol

I've included this quite unusually in the fault section purely because it may not always be wanted in a given wine. Eucalyptol is one of the few direct aspects of a local sense of place that will directly impact the flavour of our wine. The compound eucalyptol is derived primarily from eucalyptus trees; it's made volatile by heat and then absorbed into the waxy bloom of nearby grapes. Eucalyptol has an expectedly camphor, menthol and eucalyptus-like flavour which is transported into wine via the skins. Much of the research on this effect has been performed in southern Australia, where winemaking regions like Coonawarra have distinctly menthol styles of Cabernet and Shiraz and a direct correlation can be drawn in relation to the amounts of eucalyptol found in these wines versus proximity to eucalyptus groves.

Lightstrike

This increasingly common fault has arisen from the popularity of clear bottles for white and rosé wines. It's caused by light that tends towards the blue or ultraviolet end of the spectrum transforming the amino acids in wine into some awfully smelly compounds.

These include dimethyl disulphate, which in higher quantities can give off odours of damp cardboard or cabbage whilst completely blocking the presence of fruit flavours in the wine. The damage can be done in a matter of hours and is caused by direct sunlight or certain forms of fluorescent and LED lights. The best way to prevent it is simply to avoid the use of clear bottles and keep your precious bottles away from the light.

Mouse taint

This is a tricky fault that's more often found in the natural wine scene. It presents itself as a mouse-cage smelling, popcorny, rice cake and corn chip aroma. Mouse taint is a new phenomenon as faults go, and only recently have we got to the bottom of what's going on here. One of the main compounds responsible for this fault is 2-acetyl-3,4,5,6-tetrahydropyridine (ATHP) which is made by lactic acid bacteria, *Brettanomyces, Dekkera* yeasts and a few other spoilage microbes whilst in the presence of acetaldehyde (oxidized ethanol). The presence of a diverse array of microbes in natural ferments and expected lower levels of SO_2 in natural wines is the reason this fault presents itself more in these circles.

What's most intriguing is that it's often detected only after tasting, and there's good scientific reason for this. ATHP is most volatile in high pH conditions. Wines with high acidity containing ATHP may not initially present symptoms, and those with lower acidities may show a little presence on the nose before sipping. When we drink the wine, our near-neutral pH saliva mixes with the wine and lowers the acidity dramatically making ATHP more volatile and thus more detectable. The wine is already in our mouths; thus we smell ATHP's volatile aromas as they depart our sensory system, via retronasal olfaction (see page 208). It's fascinating to think how a wine has the potential to smell noticeably different in the glass compared to when ingested.

Mouse taint is the perfect little fault to finish on and leads us nicely into thinking about how it is we physically perceive wine, food and what we call flavour.

7. Flavour perception

Flavour perception

And now to the part of wine we so often take for granted: ourselves. The sommelier has long been caricatured as a snooty type, one who is expected to decipher a kumquat from grandma's apple pie in a glass of wine. Sommeliers are masters of an expertly honed craft, with finely tuned palates and many years under their belts dedicated to poring over aroma kits and sniff boxes. The rest of us, who are not sommeliers, have accumulated our sense of flavour from the plentiful years of successfully eating and drinking on this planet.

If we are all to understand wine better, we should learn how everything in the glass makes it to our brains for processing: an element of wine tasting that even those soms could do well to add to their expertise.

When describing wine after having tasted it, most fumble to find the right words, and some blame their ability to smell or taste. We are all unique and fallible wine flavour perception tools of sorts, and to a very Buddhist standard, our perception of the wine around us is aligned with an understanding of ourselves. The effect on our perception comes from within.

We often use terms such as flavour, taste and smell interchangeably when talking food and wine. 'This wine smells sweet', 'Your lamb tastes like it has rosemary in it' and so on. Flavour by scientific standards comprises taste, smell and the often-overlooked aspect of texture in food and drink. When we consume, we are taking in all the inputs from our eyes, noses, tongues and mouths to create an overview of flavour. Our sensory receptors have developed over millions of years and are primarily responsible for keeping us on this planet for as long as possible.

What's taste got to do with it?

Taste has always been a tough matter to discuss. We know that from a physiological perspective there are just five distinct tastes our mouth can detect which span the range of sweet, sour, salty, bitter and umami. But doesn't it feel like we can taste much more than this? In a way this assertion is true, water is a wonderful example of something we can taste, but we don't have a specific taste receptor for. We are all so accustomed to the taste of our own saliva that when water dilutes or washes it away, it is the absence of our saliva's natural taste which gives water its own but opposite taste. This makes the taste of water something unique to the perception of an individual, since the taste profile of our own saliva would vary from one person to the next. This is a great first step in understanding that taste perception is something unique for everyone, so if a sommelier can taste one subtle thing in a wine, it doesn't always mean you're less of a nuanced taster for not picking up on it.

Before going into the nitty gritty of our personalized taste biases in food and wine, I think it wise to cover the basics of the five tastes and how we detect them. For quite some time, we sectioned off the tongue into four sections, taste map diagrams I am sure we remember from our school days. However, this visualization of taste mapping has recently been disproved. We've discovered that the papillae that encapsulate the taste buds on our tongue do overlap quite a bit, with many areas of the tongue able to pick up all five tastes in the same location. These findings have also uncovered that taste receptors are not just limited to our tongues and also can be found across large swathes of our digestive systems.

Sweetness as a source of energy and nutrition

From an evolutionary perspective, sugars have proven invaluable as concentrated forms of sustenance. When we drink and eat sweet wine or food, the sugar molecules dissolve in our saliva and trigger our taste receptors. The specific sweet taste receptors T1R2 and T1R3 will in turn send a signal to the brain which lets us know that what's in our mouth is sweet.

Umami, associated with protein intake

Here it's the taste receptor T1R3 again which doubles up to pick up glutamate amino acids too; another receptor called the T1R1 also assists in detecting the same. Glutamate is present in proteins. Nucleotides are other umami-tasting molecules found in DNA and RNA, some of which we find in food and wine. Some of our taste receptors respond only to glutamate, and others to both glutamate and nucleotides. Ageing and fermentation of all food and drink tends to increase the concentration of these savoury tasting molecules, an effect we can taste when sampling aged soya, cheese and, in some cases, wine. *Vitis vinifera* stores glutamic acid mostly in the skins, it's one of the grape's major amino acids and if not broken down in fermentation will leave residual glutamates in a wine. Glutamates are significantly higher in many a dessert wine, where fermentation is stopped midway, leaving a little residual sugar behind, which also renders the wine sweet. These interrupted fermentation style of wines include fortified wines like Port, and even those where cooling the fermentation has been used to slow and stop the yeast from working like in Moscato d'Asti. It's also been recorded that wines which have undergone yeast autolysis will also display higher levels of glutamate.

Bitter encourages us to spit

It's TS2R proteins which act as bitter receptors on the tongue. We all have a genetic predisposition to carrying more bitter receptors than any one other, and we are all extremely well versed at recognizing bitterness as a result. To put this in perspective, compared to our ability to taste sugar, which we can detect at a threshold of around 10,000 micromoles per litre, we can taste bitterness at less than 25 micromoles per litre, making us over 400 times more sensitive to bitter than to sweet. It is no surprise to learn that evolution has played a large role in shaping this outcome. In nature, many bitter compounds, like alkaloids found in plants, are associated with poisonous or toxic effects. Over thousands of years of foraging, those with heightened sensitivity to these bitter toxins on the tongue spat and lived to pass on their genes, whilst those with lower sensitivity to bitterness would suffer the mortal consequences. Present day *Homo sapiens* have inherited this instinct from their ancient ancestors. As children we intuitively detest bitter food and drink like coffee and beer – Barolo too. Only through self-guided learning do we override the dislike of bitter to reap the benefits of caffeine and alcohol.

It's useful to remind ourselves that we were not designed to enjoy bitter and there are loads of bitter compounds found in wine, such as terpenes and phenols (anthocyanins, flavanols, flavonols). An individual's dislike of short chain bitter tannins is a very rational dislike as opposed to just a personal preference. With time that individual may come to enjoy it, with the positive psychological reinforcement of the nutritional and alcoholic benefits of the wine, but without understanding the individual's personal taste profile we cannot begin to tell people what they 'should' like.

"We are over 400 times more sensitive to bitter than to sweet."

Ever heard of a supertaster?

Supertasters are individuals whose taste buds are genetically very sensitive.

The test to catch a supertaster is in their reaction to eating Brussels sprouts. Those who find them unbearably bitter are most likely supertasters and the majority of us, who can put up with sprouts (or merely dislike them a bit), are referred to as 'medium tasters' or 'non-tasters'. Brussels sprouts, cabbage, broccoli and wine all contain a compound called phenylthiocarbamide (PTC) which presents itself as extremely bitter to the minority of the population, the 20–25% who are supertasters. PTC is detected by the bitter taste receptor TS2R38. This receptor is genetically inheritable and we all retain two copies of the gene responsible for the receptor. Supertasters have two functional copies of the gene, whilst non-tasters and medium tasters have either two nonfunctional copies or one functional and one nonfunctional copy. As a result, supertasters are far more sensitive to bitter, sweet, umami and salty tastes.

There has been good research on the topic of bitterness as recently as 2009, which proves strong links to our bitter receptors triggering the release of nitric oxide gas. This gas is an antibacterial our bodies release to ward off potential illness, an ingenious development to strengthening our bodies when eating potentially dangerous bitter foods. In the same research, supertasters tested more strongly for this bitter reaction than non-tasters, something we'd expect to see, but it also suggests that supertasters should therefore be more illness resistant.

Like all the cells in our body, our taste receptors will eventually die, and be replenished. As we age, the rate of replenishment diminishes, and we're left with fewer taste receptors. Since most of our taste receptors are bitter receptors, older individuals are less sensitive to bitter food and drink than younger tasters – a great explanation for kids not wanting to finish their veg. From a wine context, when suggesting suitable wines for guests or customers, should we now be putting more thought into the age of a person? More so perhaps than the age of the wine?

Sour, signalling spoilage, the bad and the unripe

Unlike the sweet, bitter and umami taste receptors which work as a lock and key style mechanism, communicating in a purely binary fashion with our brains, sour and salt perception work quite differently. Instead, we have tiny channels on our tongue that allow salt and acid ions to be guided through cell membranes directly into our cells. Both salt and acid ions carry an electrical charge, which in turn changes the charge of the cells they enter. Our detection of this charge is what signals to the brain whether it's salty or sour.

When we discussed what acids were (see p. 60), we discovered that an acid refers to a substance that contains hydrogen and is capable of donating a proton (H^+). Here the hydrogen ions have a positive charge, so positively charged cells on the tongue send signals to the brain that give us the perception of sourness.

"Supertasters are far more sensitive to bitter, sweet, umami and salty tastes."

Does more acidity mean more sourness?

It's a common misconception in wine that all things sour relate to high acidity (low pH), but as you might expect by now, it's not as simple as we'd like it to be. Our main measure of the acidity of a solution is the pH. Technically this is the concentration of free hydrogen or hydroxyl ions in water. However, a solution's ability to disassociate H^+ ions also affects its acidity; this is not detected using measures of pH since dissociation is an acid's 'potential' to release its H^+ ions. This potential is measured in pKa.

Our sour channels are triggered by both pH and pKa in a wine: both are measures of acidity when it comes down to it, but we don't consider the extra pKa when measuring acidity using only pH. This means that when we taste sour notes in wine, it may not necessarily always correlate to high acidity from a pH perspective.

A great practical example of this is with hydrochloric acid. It's a super strong acid and can almost entirely disassociate in water to 100 per cent, totally breaking apart into its positive hydrogen ions and its negative anion counterparts. Obviously nobody would ever drink it, but were you to do so it would taste super sour even when heavily diluted. Our old friend acetic acid (starring role in vinegar) can only disassociate to 1 per cent in water, it doesn't easily release its hydrogen ions so it measures as a weaker pH acid, yet is also super sour to taste.

Our sour channels are nature's inbuilt mechanisms to detect when food has gone bad if turned sour by microbes, or when it's not quite ripe and ready to eat. Our sourness perception is a remarkable system for attaining maximum nutrition whilst avoiding the risk of illness. Higher levels of sourness will encourage most animals to spit, however like bitterness, most humans through our varied experiences have come to embrace the sour.

Salt as a valuable source of sodium

Sodium is a vital part of our diet and much needed for a healthy nervous system, muscle contractions and keeping a balance of the minerals in our bodies. Salt is made up of the metal sodium (Na^+) and the halide chlorine (Cl^-) to form NaCl. Salt has no taste until dissolved in water, and when it hits our tongues, the positively charged sodium and negatively charged chlorine molecules get to work on sliding through the appropriate channels on the tongue to indicate to the brain there's salt in the vicinity.

Our salt channels are technically not salt channels but NaCl channels. There are other salts that exist in nature, yet we can only taste the salt NaCl since our salt detecting channels on the tongue are so small that they can only let the tiny (Na^+) and (Cl^-) ions in, whilst keeping larger ions and particles out.

As a reminder, salt levels in wine are at an imperceivable level. Salt cannot be smelled either. Any impressions of salinity on the nose or to the taste in wine would be the result of other factors that we may associate with saltiness and are not the detection of actual salt.

In science, taste is viewed as the interplay between genetic and environmental factors. The make-up of our taste receptors is inherited from our parents, though environment can dramatically change our perception when it comes to taste. It's for this reason that good wine is an acquired taste which requires adequate training and experience to recognize, whilst accepting that great tasting wine for one person may not be so enjoyable for another. It doesn't bode well for us to think of wine as purely a subjective matter, since that would render the subject of learning about wine quite redundant – a somewhat nihilistic approach to wine. This element of subjectivity may be the reason that instead of learning about wine itself, most wine education programmes focus on what we should expect in wines from a certain region, or typicity. What is then said to make these wines good or bad is how tightly they fit into those expectations of typicity. This flips the whole concept of wine being a representation of its terroir on its head, and alludes to wine being made to fit the parameters of what we've learned it should taste like rather than having tastes specifically imbued by that place.

And spice? If we can feel it on our tongues, why isn't there a taste receptor for it?

The term 'spice' comes from the Latin for 'special wares', a very broad derivation that in no way refers to a style of ingredient. So let's first define what we mean by spice: are we talking fiery pepper spice? Or cloves and aniseed? For the sake of simplicity let's speak about spice only in reference to that fiery sensation some spices leave on our tongue. How can we sense it if there aren't any taste receptors for it?

Chemesthesis is the answer. This describes the phenomenon of plant chemicals triggering sensations on the tongue. The effects range from creating the perception of heat, like capsaicin in chilli peppers, which causes physical pain to sensitive tissues like the tongue, or having the opposite, cooling effect that menthol gives. In the case of menthol, its chemesthetic properties block the calcium current along the nerves responsible for temperature detection. More chemesthetic ingredients include Sichuan peppers, which have an anaesthetic effect, and fizzy wines that create a physical tingle. They are all physical and chemical sensations created on the tongue, not technically tastes, though part of a broader chemesthetic sensation.

The chilli compound capsaicin can also be present to small degrees in some wines and can be responsible for any chemesthesis-related spiciness in wine. It's worth noting that things like rotundone which also present as spicy, black pepper, and woody notes do not have a chemesthetic effect, we are simply tasting and smelling these.

How smell works

If we can only taste these five different components, how do we then tell the difference between a strawberry and a cherry? This is where the previous definition of flavour comes in. We would all struggle to differentiate using just taste without the assistance of the other components of flavour, smell and texture. It's even harder with wine, the texture of fruit is eliminated entirely, it becomes wine textured as opposed to having the mouthfeel of an actual strawberry. It's a step harder to define as we're wholly reliant on our ability to smell the strawberry to define it. You can experiment at home with puréed fruit: let's use the strawberry and cherry example. Hold your nose and try and tell the difference between the two purées. By discarding the texture and some of the smell element, you will just be relying on levels of sweet, salty, sour, bitter and umami, which renders us all pretty useless as tasters. Different things may all taste sweet, but smell is the thing that differentiates them all.

Unlike taste, where many aspects are coded into our DNA, smell is learned. We have around 400 different odour receptors located in the olfactory mucosa under the roof of our nasal cavity. Information from the olfactory system (the sensory system for smell) is sent a very short distance to the brain, where these 400-odd different inputs are shaped into more than 10,000 distinct smells that we teach ourselves to recognize. Our perception of smell is essentially analogue, where taste is a simpler binary process with built-in genetic biases. In many ways this does correlate to how professionals in the wine industry develop their palates, years of sniffing boxes to train oneself in putting a name to a smell input is exactly how our olfactory system learns, and each complex sensory input needs a name.

Picture the scene: we sit around a tasting table and all take a whiff of wine. The one proclaimed expert in the room picks up pencil shavings, forest floor and black gooseberries, and the rest of us cautiously nod and murmur in agreement. We covertly question ourselves on why we can't pick up any of that. Let's stop to think. Is it because we can't physically smell what the expert smells? Or because we can't put a name to the smell that we're all smelling? In most cases it's probably the latter. When someone trains their palate, rarely does their sense of smell physically develop, but their ability to put a word to those smell inputs improves. This thought should give us hope, I find that acknowledging and overcoming mental hurdles is far easier than tackling the physical ones.

As you tilt a glass of wine towards your nose, the most volatile molecules lift to the rim of the glass and evaporate into the headspace. A waft of aroma molecules travels through our nostrils to the olfactory epithelium, where thin hair-like formations called cilia poke their heads through a layer of mucus, which traps the odour molecules. The odour molecules then dissolve and bind with the olfactory receptor cells that send a signal to the olfactory bulb tucked away just underneath the brain's frontal lobe. The next step is to take this information to the piriform cortex where these odour pictures are formed for us to decipher. This entire process is referred to as orthonasal detection. New odours we encounter leave a mark in our memory, and so our recognition of smells develops.

Smell is created in the brain and projected like a picture, and some can communicate this picture better than others. A scent picture gives us a personal emotional stimulus such as pleasure or disgust and is then stored away. Often at the bars a guest can tell you if they like or dislike a wine but can't explain why or what it is about it that makes them react that way. This communication aspect of smell is viewed as a skill; to reduce it somewhat, it's simply describing the scent picture to others in an accurate way. Very often wine lovers with a cynical disposition will mention they are only smelling what they smell because they were told they would by the sommelier. It's technically not correct, but somewhat true, like a visit to the palm reader. We are able to detect the many scent pictures in our glass of wine but can't put a name to it, and when the expert pulls a detailed account of any one of those scent pictures from out of the hat, we instantly recognize it. It was not a trick of the mind, but an assisting nudge where the eureka moment was in aiding us to recognize and describe the smell, not the discovery of the smell itself.

This recognition system is a synthetic process. As we learned from the different components of a grape, there is no one 'grape odour'. There are many. When these numerous volatile aromas reach our olfactory system, we decipher from the multiple inputs that this is a grape we are smelling. Psychophysical studies have shown compelling evidence that the perception of different volatile aroma mixtures is not simply the sum of all individual descriptors, since they form completely different odour images in our brain regardless of how chemically similar the make-up of the input. This occurs for all aroma mixtures that contain more than four volatile components.

Our brain will process these four or more aromas as a completely different final image irrespective of its individual components. Changing one of these individual components will likely create a completely different odour image in our minds. This explains why we perceive such a spectrum of smells from single ingredients like wine and coffee after processing. They carry such numerous mixtures of volatile aromas and smells which were never there in their base ingredient in the first place. It's the reason we perceive gooseberry in one wine, lemons in the other, and ripe mango in the next, all of which may be from the same region, vintage or variety. Just one or two individual volatile aroma input changes will completely change our final image of it. In the field of psychophysics, this phenomenon is known as synthetic processing.

Studies have demonstrated that almost 90 per cent of our flavour experience will come from the olfactory system: smell is king. That statistic may seem hard to believe so let's look at a couple of examples. A sad real-world case study on flavour is related by author Brillat-Savarin, who reports in his celebrated work *Physiologie du goût* (The Physiology of Taste) that he met with a messenger in Amsterdam whose tongue had been cut out by the Algerians as punishment for attempting to escape their clutches. He reported: 'Persons born without a tongue, or whose tongue has been cut out, are not completely deprived of the sensation of taste.' In the case of the messenger, little loss of flavour perception was noted. His book is a morbid yet fascinating documentation of how our flavour systems all work together. Inversely, those of us who have suffered from a loss of sense of smell through COVID, or have even just had a heavy cold, will know all too well how it felt as if we couldn't taste anything. Yes, you heard it right, our loss of smell inhibits our perception of what we taste. That statistic is looking a little more convincing now.

From an evolutionary standpoint, it makes total sense that smell should form the larger and more complex part of flavour detection for us and most animals, much like sight creates a way of distinguishing potential dangers before it's too late. It's far more effective for us to smell if meat is rotten before we put it in our mouths, since at that point it may well be too late! Even today we humans are influenced in the most extreme ways by our subconscious processing of smell. Subconscious smell decisions guide us towards which partner to choose, which jobs to work in, and smell acts as a superpower for pregnant mothers, whose olfactory system is turned up a notch to help them protect their unborn babies.

What are smells?

We first encountered volatile aromas in Chapter 4; here we'll look at them in more detail. Volatile aromas are carbon chains light enough to float in air to reach us. These carbon chains fall under four main chemical families: acids, alcohols, aldehydes and hydrocarbons. Larger, 'heavier' chains may get broken into smaller constituents by metabolism or other physical processes like heating, which can introduce new, 'lighter', volatile fragments into the air, thus creating new aromas. We see this every day: for example, a soup smells stronger when heated, or introducing acids, in the form of citrus, to make ceviche, completely transforms a raw fish dish. In each example we're chopping down long carbon chains into smaller and more volatile fragments.

The reason we chill certain wines more than others is all about aromatics. In fancier restaurants, some white wines will be served warmer than others. These tend to be the fuller and less aromatic styles of white wine, such as your typical white burgundy, which benefit from that extra bit of warmth to make the volatile aroma molecules more volatile, assisting the wine in being physically lifted out of the glass. Aromatic styles of wine, such as Sauvignon Blancs and Rieslings, don't need the help so can be chilled to lower temperatures and still display enough of their volatile aromas. Those same wine enthused restaurants will often choose to assign different glassware to wines expressing different aromatic potencies. The less chilled whites that needed that extra nudge to lift the aromatics out of the glass could also benefit from wider glasses. The aptly named Burgundy glass may do the job as it has a phenomenally wide waist. The wider part of the glass is the level the wine should be poured to in order to best benefit from exposing the most amount of wine surface area to the atmosphere. The oxygen will start picking away at the surface and reacting with the wine very slowly. Decanting a wine serves the same purpose as micro-oxygenation in the glass, exposing and reacting new layers of wine to oxygen, and urging volatile aromas to be lifted. This reaction is colloquialized as 'opening up a wine' in sommelier lingo.

Other less obvious environmental factors can also change the ways molecules behave. All aroma molecules will behave differently in solvents like water, alcohol, fat or air depending on their particular physical make-up. Hydrophobic molecules hate water, and much prefer dissolving into fats, whilst hydrophilic aroma molecules love being in water. Ethanol has only partially hydrophobic properties which explains why the aromas found in wine and spirits remain there, even with alcohol present. It also explains why food dishes smell entirely different when boiled in water, fried in oil, grilled in open air or macerated in alcohol. A fun experiment to test this effect: grab a tumbler of whisky and inhale the aroma. Now, add an ice cube or drop of water, and smell the whisky again. The whisky would be said to completely 'open up', it's far more expressive with the water or ice applied. We might think the ice would chill and 'close' the whisky aromas making them less volatile and less aromatic, but it does the opposite. The water provokes new smells, and more of them. The addition of more water to a high alcohol mixture encourages the hydrophobic volatile aromas to excite and want to leave the glass: they were once quite happily sitting dissolved in alcohol, and then are made volatile by the addition of water.

This also implies that wines of differing alcohol levels will alter their aromatic profiles based on their final abv and the resultant make-up of alcohol-loving and hating aromas in the wine. It's very tricky to say broadly what specific aromas are lost or kept, mainly since our perception of aroma is a synthetic one, with any single change of input completely changing our final aroma image. Just a few degrees change in alcohol levels (percentage abv) has the knock-on effect of altering the smells we pick up in our final wine in a sizeable and rather unpredictable way.

Further to this, every aroma molecule that reaches our olfactory systems will have a unique odour detection threshold – the lowest concentration of a compound that can be perceived by a human's sense of smell. If there's a volatile aroma in the air that we can measure, but there's not a high enough concentration of it, we won't be able to smell it. And each specific odour will have a different concentration threshold that is unique to each person.

"A few degrees change in alcohol levels has the effect of altering the smells we pick up in our wine."

Smell aerodynamics

Orthonasal versus retronasal olfaction

To add another dimension to how smell works, there are marked differences in what we smell based on how we do it. Orthonasal smell is the term for taking a smell in through our nostrils, i.e. how most would naturally imagine smell operates. Volatile aromas leave our glass of wine, enter the air, hit the olfactory system when we breathe in, and then depart with the air from our lungs as we breathe out. This is how we've regarded the airflow of smell thus far, and when smelling a glass of wine that is exactly how it works. This mixture of air and volatile aromatics going in through our nose is high in oxygen and at room temperature, hence the aroma profile accommodates molecules that sit well in this environment.

Retronasal smell occurs after we swallow our wine or food. It begins with orthonasal olfaction, then after we swallow, a little bit of time in our warm bodies superheats the gulp. This gentle warming of the wine by our mouths makes the aroma compounds more volatile. The base aroma notes which we could not initially perceive suddenly become volatile and join the airstream which leaves our bodies on the breath out. These superheated volatile aromas differ quite a bit from those aromas that once came in, since our breath out is carbon dioxide rich and invariably warmer. On exit, much of these gases will pass back through the olfactory system, building an aroma image from these alternative components from the same stimuli. It all happens almost instantaneously and our piriform cortex will build a picture of smell based on both orthonasal and retronasal data. This gives us a much fuller and more accurate picture of what it is we are eating or drinking based on a wider spectrum of volatile aromas that are released both coming in and going out.

The incessant gurgling and swilling of wine in the mouth, a display by sommeliers and wine tasters around the world, is now starting to make sense. Wine mastication benefits from the trifecta of washing the wine around as much of the tongue as possible to pick up more taste, introducing more oxygen to the wine to release more volatile aromatics, and forcing an element of retronasal smell by superheating the wine in your mouth. For those wanting to build a bigger picture of the wine we smell and taste, mouth swilling is an essential piece of kit in the armoury.

Sniff velocity

Research has shown us there are marked differences between normal, rapid and forced smelling. Odour molecules all have differing rates of sorption into the mucus of our olfactory systems. Generally, weightier volatile aromatic molecules will require more time and slower sniff velocities to settle into the mucus. It's like passing a hoover over lighter bits of dirt that will be sucked up quickly, whereas heavier dirt components will need a little more time spent on them in order to get hoovered up. Experiments with creamy Chardonnays show that the weighty and buttery aromatic compound diacetyl is most notable when slow sniffing rather than with forced sniffing trials. It's fun to know, but is it useful? Well, if we're looking for specific aromatics in a wine then yes it is helpful to know that harder sniffing may not always reap the rewards we want. Some middle ground of shallow rapid sniffs as well as slower longer ones may be the trade-off we need.

Nasal cycles

I often bump into the same wine sniffers from one trade tasting to the next: it's fun to observe other professional tasters who, like me, all have a unique style and method to documenting wines. Many lean the glass to one side of their nose, quite possibly what they believe may be their beefier, better-performing nostril, and take a deep sniff in. It isn't necessarily best practice to rely on one nostril for all of our smelling, since our nasal cavities on average exchange congestion from one nostril to the other every 90 minutes or so. Unless you're ill and both are blocked, this'll mean that one free nasal cavity is always available when we need it, but it won't always be the same one – if we are always sniffing with the same nostril, we may catch it on its 90-minute break.

Psychological state

We have all heard tales of our environment changing the enjoyment of wine. We reminisce on times spent on a beach in Greece, sipping on Assyrtiko. But, when we take the same bottle home to rainy London, it just doesn't taste quite as good. How much science is there in this?

The closest study which touched on this topic was in 1972 with rats. Hungry rodents were compared to their sated counterparts in studies led by French physiologist Jeanne Pager, who measured the impulse responses of the mitral cells connected to orthonasal food detection. A clear correlation was found, with significantly higher orthonasal responses documented in the hungry rats than the sated ones. It seems obvious because we would also react in the same way. This experiment is important as it proves the connection between our psychological state and our observation of the world around us, from one moment to the next. Knowing this now makes food and wine pairing recommendations a very tricky business, plus shows us the extremity of how subjective our likes and dislikes are, in relation to our internal state.

Saliva makes a difference

There is yet another interesting layer of complexity which arises from our saliva. The make-up of saliva for each of us is unique and ever changing, and also has a notable impact on the taste and smell of wine. Saliva functions as a lubricant, helping in digestion and speech, and acting as a protective layer against a multitude of threats such as acids. We can look at saliva impacting flavour in a few ways, firstly as a flavour blocker. A great experiment to try yourself is by comparing flat sparkling wine to its bubbly counterpart. A flat Champagne is notably sweeter tasting than the freshly opened bottle; this applies to fizzy soft drinks too. In the bubbly instance, research has illustrated that our saliva prevents the carbon dioxide bubbles from travelling from the tongue to the back palate. The bubbles settle for a short period in our saliva, and get backed up as more Champagne is applied, forming a bubble barrier that blocks the sweet components of the wine from reaching our sweet receptors. The non-fizzy morning-after Champagne doesn't have this effect and we can taste the sweetness in the wine for what it is.

The second instance of saliva affecting the flavour of wine is in the physical make-up of our saliva. It's full of proteins like mucins, bacteria and enzymes, which interact with the wine to release certain aromatic volatile components.

One fascinating experiment used Sauvignon Blanc to look at the effect of saliva on thiols and flavour perception. The experiment concluded that enzymes in our saliva initiated the release of the exotic fruit/cat pee aromatics from Sauvignon Blanc. The longer the wine sat in a person's mouth, the more their salivary enzymes released these aromatics from the wine and the more aromas they perceived. Some individuals will carry more of these thiol-releasing enzymes, whilst others may produce a higher flow rate of saliva, all factors of course will have a marked impact on our personal perception of wine.

Thirdly, we look at the flow rate of saliva itself in relation to retronasal olfaction. Those who salivate more than others will be engaging more retronasal aromas than others, as well as exposing more aromatic releasing enzymes into their wines than others. It's also worth noting that we all individually salivate more or less for all sorts of reasons, be it diet, age, illness, mood or hunger.

Are genetics involved in smell at all?

Yes, is the short answer. There are smells which we are predisposed to being more sensitive to than others. We have no odour receptors for water (H_2O), carbon dioxide (CO_2), or nitrous oxide (N_2O), which means we have no ability to smell these things. We do contrastingly have hypersensitivity to the eggy smelling hydrogen sulphide (H_2S), detecting the smallest of traces present in the air. In nature, sulphurous smells are present around volcanoes, hot springs and rotting organic matter. Their sources can be irritants or even fatal. It's no doubt an evolutionary advantage for those with higher sensitivities in staying alive. Smell thresholds are fundamental to the survival of many animal species – dogs possess a low odour threshold for carbolic acids, present in their natural prey; mice have heightened sensitivity to smelling their predators and camels have hyper-low geosmin thresholds, enabling them to detect damp earth from a distance of 50 miles! We're told that dogs have a much better overall sense of smell than us, which is true, yet humans outperform dogs when detecting aldehydes – present in fruits and flowers, and have a lower threshold for geosmin in soil than sharks do for blood in water!

In terms of odour thresholds, genetically our species follows a broad correlating spectrum. For hydrogen sulphide or geosmin, most of us will have exceedingly low odour thresholds, enabling us to smell the tiniest of concentrations, but we each have varying thresholds for each odour. The most extreme examples swing so far that some of us can't smell any of one aroma, regardless of the concentrations present in the air. This trait is called anosmia in relation to olfaction, and ageusia in regard to a change or loss of taste. Anosmia is more common than one would think. We've already discovered that just below 25 per cent of us have anosmia to rotundone, the peppery volatile aroma found in Syrah, Grüner Veltliner and Malbecs. Many of us just can't smell it, regardless of how much there is in a wine. This one fact alone should help us rethink the theatrics of listing all perceived smells in a wine as a feat of wine prowess. We're coming to understand that smell is a unique perception, like art. When we look at the Mona Lisa, we all see the same picture but describe it in all sorts of different ways. It would be comical if we gazed upon the Mona Lisa and in a show of artistic expertise described all the emotions we felt as a result of looking at it, yet we partake in these displays quite regularly in wine.

There are numerous arguments which play to the idea of an individual's genetic disposition for wine tasting. After reading the last few pages it may indeed appear to be true. More receptors, lower odour thresholds and better mechanisms for smell will of course assist us in building a more accurate picture of a smell but it's our brain that does all the magic. In the beginnings of humankind, our close competitors in the animal kingdom all had better senses of smell, hearing and sight than we did. Even with this diminished natural advantage modern day *Homo sapiens* can compose music and create art, we play and do more in a creative context with the relatively poor senses we have, than most. I argue that our ability to perform well in wine, love or life does not rely on our genetic make-up, rather we need to focus on what our brains enable us to do with the inputs and opportunities we have.

We've discovered that it's indeed true that our physical ability to smell differs as individuals: research shows that 30 per cent of our olfactory receptors do differ from one person to the next. Because of this we will all perceive marginal differences when smelling the same thing.

Texture as a component of flavour

Texture is a key factor in our enjoyment of wine and food. We have more touch receptors in the epithelial layer at the front of our tongue and mouth than any other one place in our bodies, and they're responsible not just for determining the size and shape of objects but also texture, temperature, pressure and pain. This is another crucial gatekeeping measure to keep potential dangers out rather than in.

Somatosensation is the term we use for the body's ability to touch and feel. Though retronasal smell forms the majority of our image of wine flavour, it is the somatosensory texture and feel on the tongue which is often a subconscious factor in us determining whether we like it or not. At the bars, the trickiest part of a guest's experience is most often articulating why they do or don't like a particular wine. With red wines especially the feel of tannins can be an overriding factor, regardless of how tasty the wine is. For an individual who's not aware of what tannins are or what that sensation is, it's nearly impossible to explain why they may not like the wine. Texture is all too often the deciding factor in a glass of red, even if it only builds a fraction of our flavour picture.

Our somatosensation system can process information on light touch, location, direction of movement, force, pressure, temperature and pain. These variables combined build an image of what we call the mouthfeel of a wine and there are more than just a few components of wine that contribute to this. Tannins are the obvious ones, those grabby flavanols are something we physically feel binding onto the proteins of our tongue, mouth and saliva. The binding and precipitating of our saliva gives the added sensation of increasing friction between mouth surfaces, giving a drying sensation. If we think back to Chapter 4, we'll remember that the shorter chained tannins are the more aggressive and noticeably grippy, whilst longer chained tannins feel velvety and smooth. Factors such as the part of the grape the tannins come from, what they've already polymerized with and the age of a wine will all have an influence on how long or short these grippy chains of tannins are.

Thicker components of the wine also play a huge role in the feel and viscosity of a wine. Levels of sugar, alcohol and glycerol all add thickness to the liquid. The more there is of any one of these components in a wine, the thicker the wine will feel, which may be likened to the difference in sensation between skimmed and full-fat milk (although there the thickening factor is fats). Thicker wines include the sweet, those with high alcohol and the carbonically macerated, but our enjoyment of viscous wines is a purely subjective notion. A quick-thinking subjective texture judgement from any taster will influence their enjoyment of the wine as a whole, regardless of how incredible it may taste or smell.

Can the look of a wine influence flavour perception?

Our species relies on problem solving, pattern making and remembering learned biases to prevent the same issues occurring time and time again. These skills have led us to the top of the animal hierarchy and many of these prejudices still play a part in our decision making. A remarkable study, 'The Colour of Odours', was published in 2001 where 54 undergraduates from the Faculty of Oenology from the University of Bordeaux were given two wines over two sittings. The first two wines were a white (either Sauvignon Blanc or Semillon) and a red (Cabernet Sauvignon or Merlot). The participants were then asked to describe the flavour of the wines. Their lexicon was then divided into common descriptors for white and red wines. In the second sitting the white wine had a neutral anthocyanin dye applied to it making it look like a red wine. The wine students then performed the same exercise, and the descriptors were noted and separated accordingly. It's little surprise to the cynic that dwells in all of us that the results showed a heavy shift towards flavour descriptors for red wines in the fake red, dye-tinted white wine trial. The observed phenomenon is a real perceptual illusion which shows that our sensory and cognitive processes are mostly influenced on the look of wine as opposed to taste or smell.

We all conjure assumptions to assist our senses. Our eyes begin to filter down options first, then our noses and mouths get a look in. The phenomenon discovered in 'The Colour of Odours' experiment is similar to another well-known perceptual illusion: the size and weight illusion. Here, the smaller of two equal weighted objects is almost always deemed heavier.

This visual prejudice is called upon to summon the lexicon required to describe a wine, or at least streamline the options. I must also add that knowing the expected characteristics of a grape, or characteristic expectations of a region will of course add more guided ammunition to our wine vocabulary when describing a wine we know, well before our taste and smell gets involved. I sometimes compare these tricks to those of a palm reader, which may seem extreme, but I don't think it is way off the mark when you consider how much subjectivity is involved in smell, how individual taste is and how much sight has proven to give us an overriding predisposition to smell what we expect, as opposed to what's actually in the glass.

I've lost count of the job interviews I've held for sommelier positions at the bars where candidates have boasted about possessing a phenomenal palate. How and why does this help anyone? Should we just tell guests what they should be tasting, and hope they're impressed? Or are we hoping they'll trust our expert palate more than their own? Is it our ability to tell people what we're smelling in a wine that sells it, or the theatre and performance of buying wine in nice restaurants that we enjoy and actively participate in?

The teams at both bars do their utmost to steer clear of the subjectivity in taste and flavour of wine for these very reasons, though the greater wine world still gets caught up in separating the different components of wine into flavours. It's what is taught, how it's sold, and quite often how notes are taken when wine is judged. Whoever walked into a wine shop and asked for help in choosing a bottle that tastes like strawberries? Yet, inversely when many sommeliers sell a young Pinot Noir to a customer, they insist on telling them it smells of strawberries.

If we as a people want to understand more about food and drink, let's make a start by looking beyond the theatre of flavour. We may add a flourish of romanticism to tales of wine, but what do we learn from that? Wine bottle labelling often reels off flavours to expect, historical aspects and the geology of a wine region. We're rarely given pertinent information like time on skins, how long a wine has spent ageing and in what vessel, whether it has been made using commercial yeast and the relative proportions of grapes used in a blend. It appears that in the limited space we have on a bottle of wine, even if specifications that describe how a wine has been made are fundamental to understanding what's in the bottle and deciding if we may or may not like a wine, it's the romance that sells. If we're ever to make our own decisions on wine as a consumer or professional, surely these specs should make up the basic information for the back label of a bottle.

The language of wine

Describing in words the wine we drink is so often the hardest part of a tasting experience. We possess the words, we taste the wine, yet making connections between the words and what we drink is a real challenge for all but the most seasoned of wine drinkers. Scent boxes are used in sommelier training to develop these neurological connections and do assist us in learning how to communicate an olfactory input. Neurological research on this issue of describing flavours relates it to the distance between the areas of our brain responsible for the cortical processing of language – Wernicke's and Broca's areas – and the olfactory system that's situated at the front of our brain. In brain measurement terms they're pretty far apart. The physical distance between parts of the brain seems like an archaic way to look at these interactions, however correlations have been formed between the general intelligence of animals and their information processing capacity (IPC). IPC depends on the quantity of cortical neurons, interneuronal distance and axonal conduction velocity. Studies in schizophrenia have demonstrated that brain connectivity strength lessens as the distance between parts of the brain increase. It appears that our brains are hardwired to be rubbish at explaining flavours in wine!

Over many practised years delivering wine education and tastings I must anecdotally note that often English lacks precision when it comes to finding the words

---

"Making connections between the words and what we drink is a real challenge for all but the most seasoned of wine drinkers."

to explain wine. Terms like 'tannins' aren't often understood, 'dryness' has multiple meanings, and 'sweet' is used imprecisely to describe fruit aromas as opposed to sweetness. English allows a lot of wiggle room for us to use the same words for multiple purposes. This freedom makes the language versatile and has birthed awesome poetry and arguably the best comedy in the world but makes conveying to others what we are tasting more difficult. Other, more precise languages, like Japanese or German, have a word for everything and the feeling of tannins is fixed and known, so explaining wine flavours comes a lot more easily and can be done more concisely.

The science of memory and flavour

As we wrap up our wine journey, what better subject to finish on than memory. We've covered a lot, some of it very technical, but just how much of it will we remember?

I've had the joy of working with and developing some incredibly talented, up-and-coming wine people in London. Their ability to remember the feel and profile of a wine we've trained on is exemplary, but this skill should in theory come quite naturally to most of us. Even if we can't explain the precise flavour in words, the nostalgic smell of Grandma's cooking, the mud we played in as a kid, dad's shed or specific and unusual spices from a childhood friend's house, these aromas are engrained firmly in our minds. Decades later, when the same smell wafts past, childhood memories erupt and remind us of the place we smelled it last. It happens without us even trying, without our control. The nostalgia of flavour is an automatic response at which we are all naturally adept.

Unlike the cortical language processing areas of the brain, our olfactory system is right in the middle of the part of our brain associated with memory. The olfactory nerve is the shortest of all the cranial nerves, hence information is delivered swiftly. Further to this, the limbic system, which is the area of the brain involved in emotional processing, encapsulates the olfactory bulb. Our smell centre sits within our emotional processing system and as a result emotions and smells are incredibly closely intertwined. This further influences our personal biases in wine preference. We may well decide on wines we like or don't like based on a decade's worth of good or bad emotional experiences linked with similar smells – just how is a sommelier to know? Surely a pre-dinner therapy session isn't the next stage of evolution for the role of sommelier.

The amygdala is where emotional processing is managed, and the olfactory system and the amygdala are constantly talking to each other. The olfactory bulb's next closest buddy is the hippocampus, the area of the brain responsible for memory and cognition. These connections are direct and the shortest in the brain; as a result, smell, emotion and memory are all tightly bound together for each of us. It's common knowledge that memory and emotion are closely linked, though smell is often left out of this trio.

When we smell a brand-new stimulus, our amygdala and hippocampus quickly analyse and store it. Experiments have shown that when these new smells are presented again, we can quickly reference when we last smelled them. Experience in smelling is key to developing our bank of smells and our ability to recall them. Experimentally, women have always outperformed men in smell tests at both higher and weaker odour thresholds. This is often speculated to be a societal development – it's been recorded that in most countries women will spend more time working with food and undertaking other tasks involving odour identification, meaning women are exposed to more smells than men in their lifetimes. Other research has shown that even at an early stage, female babies will display more interest in olfactory cues. Another theory around this bias speculates on data showing that women possess more negative amygdala associations than their male counterparts for the same smell stimuli. This negative association spike would translate to better memory storage capabilities when linking to the hippocampus nearby.

Conclusion

On finishing this read you are now much better equipped to go forth and challenge the world of wine. You now know more about this 500-billion-dollar industry than most. You've developed firm building blocks to make better choices, ask useful questions and further develop your understanding of wine. With your newfound understanding of cation exchange you'll be able to probe what you're told about soil and terroir at a tasting or at the table; I challenge you to try to wrap your head around whether these connections make scientific sense. It's the only way we in wine will develop a more nuanced and accurate language around the drink we sell, which in turn will nurture a more clued-up consumer base. Knowing your sand from your silt, and how soil type impacts crops, vines and wines also sets you apart in being able to separate the sales spiel from sound scientific fact.

Quality is a subjective notion, and with wine it's often hard to know where best to place our measure of value. Learning about conventional farming and the changes that occurred after the two world wars in the way we grow and scale our food production have completely shifted where I place my own personal gauge of quality. I often ask myself when drinking a wine I enjoy, just how have they achieved that? If I delve into the rabbit hole of farming and production and discover extensive use of harmful synthetic pesticides, or that copper and the like have been applied, it has the direct effect of lowering the bar of quality, in my mind. I hope that gaining an understanding of farming, organics, regenerative practices and their respective pros and cons will aid in guiding you to where you place your own set of values.

Having absorbed this book, you are at a stage where, like an oenological version of Neo in the *Matrix*, you can look at a wine on a list or shop shelf and start to decode it. You can begin to predict the drinking experience from knowing the type of climate it heralds from. You'll know whether to expect a full-bodied, high-acidity or low-alcohol wine. It's in that intimate understanding of a grape's make-up, and how the climate influences it, that we have the power to accurately predict the style of a wine. And on those occasions where we forecast incorrectly, more insight is to be gained in the failure. This engineering-led approach to understanding our grapes is key to remembering and understanding more in wine, and at a quicker pace.

Microbes are the living engines of life, and our understanding of the role they play in wine both for better and for worse is crucial to our understanding of fermentation as a whole, and the risks that lie ahead in the making of wine. They justify the means to some of the interventions we make in production, as well as illustrating the complexities around what makes a wine taste so different from the grapes it is made from. Humans have known how to harness these microbes for quite some time now. In the winery our ability to bring order to natural processes has permitted us to mass produce wine for a global market. This in turn has led to the demand for a consistent, long-term product, fit for supermarket shelves. We are now obliged to rely on those same technological advances.

What matters most is the subject we dwell on the least, our perception. Learning that our personalized flavour likes and dislikes are based on both genetic and learned biases, all of which are subject to momentary changes, could lead us to believe the exercise of wine tasting is pointless. It makes our own assessment of what we like or dislike, whether it's a good or a bad wine, completely subjective. I personally derive my own enjoyment in wine from the learning: flavours are fleeting emotional engagements, and I find the real contentment comes from understanding and appreciating.

For quite some time, the science has lagged behind our understanding of wine. We are now stepping into a new age of discovery and appreciation. This book however is not just about wine, these topics are far reaching and relate to how we eat, drink, farm and cook. This is the book that I and many others looked for when starting our journey in wine, and I hope these discoveries serve you well.

Glossary

Acacia gum: Natural gum originally consisting of the hardened sap of two species of the Acacia tree, *Senegalia senegal* and *Vachellia seyal*.

***Acetobacter*:** Bacteria that oxidize organic compounds to acetic acid, as in vinegar formation.

Adsorption: Surface phenomenon that occurs when a gas or liquid solute bonds to the surface of a solid or liquid, forming a molecular or atomic film.

Agrochemical: Chemical used in agriculture, such as a pesticide or a fertilizer.

***Alternaria alternata*:** Fungus responsible for leaf spots, rots and blights on many plant parts, and other diseases.

Amino acids: Molecules that combine to form proteins. Amino acids and proteins are the building blocks of life.

Amygdala: Region of the brain primarily associated with emotional processes.

Anion: Negatively charged ion.

Anthocyanins: Class of colourful water-soluble flavonoids widely present in fruits and vegetables.

Anthroposophy: Formal educational, therapeutic and creative system established by Rudolf Steiner, seeking to use mainly natural means to optimize physical and mental health and well-being.

Arboretum: Botanical garden devoted to trees.

Axonal conduction velocity: Speed at which an action potential travels along an axon or nerve.

Barolo: Dry, full-bodied red wine from the Piedmont region of Italy.

Basalt: Dark fine-grained volcanic rock that sometimes displays a columnar structure.

Beaucastel: Château de Beaucastel is a winery located in the southern part of the Rhône valley in France. It is primarily noted for its Châteauneuf du Pape wines.

Bedrock: Solid area of rock in the ground that supports the earth above it.

Benzenoid: Class of organic compounds with at least one benzene ring. Benzenoids have increased stability due to resonance in the benzene rings.

Biosynthesis: Process by which living things use chemical reactions to create products useful for cellular metabolism.

Bisulphite: Salt or ester of sulphurous acid containing the monovalent group $-HSO_3$ or the ion HSO_3^-.

Broca's area: Located in the left inferior frontal gyrus of the brain, this area is responsible for speech production and articulation.

Calcareous: Mostly or partly composed of calcium carbonate.

Camphor: Strong smelling powder, originally derived from the bark and wood of the camphor tree. Today, most camphor is synthetic. Think Vicks VapoRub.

Cane: Part of a vine that is neither woody nor a green shoot. Usually one-year-old vine growth.

Canopy: Parts of the vine visible above ground – the trunk, cordon, stems, leaves, flowers and fruit.

Carbon chain: Series of carbon atoms linked together by covalent bonds, the backbone of many organic molecules.

Carotenoids: Class of mainly yellow, orange or red fat-soluble pigments, including carotene, which give colour to plant parts such as ripe tomatoes and autumn leaves.

Chablis: Dry white Burgundy wine, made from the Chardonnay grape, from Chablis in north-central France.

Charge: The physical property of matter which causes it to experience a force when placed in an electromagnetic field. Electric charge can be positive or negative.

Chirality: The characteristic of an object that cannot be superimposed on its mirror image. A widespread phenomenon in nature.

Chlorophyll: Green pigment found in plants, algae and cyanobacteria which gives them their colour and helps them make food through photosynthesis.

Chlorosis: Another term for yellowing in plants.

Co-enzymes: Organic compounds required by many enzymes for catalytic activity. Often vitamins, or derivatives of vitamins, they can act as catalysts in the absence of enzymes, but not so effectively as they do in conjunction with an enzyme.

Collagen: The main structural protein found in skin and other connective tissues.

Covalent bond: A stable chemical bond that involves the sharing of electrons to form electron pairs between atoms.

Cryoextraction: Process by which grapes are frozen and then pressed.

Cultured bacteria: Laboratory-grown bacteria.

Cyanobacteria: A division of photosynthetic bacteria.

Decarboxylation: Chemical reaction that removes a carboxyl group and releases carbon dioxide (CO_2).

***Dekkera*:** The asexual stage (anamorph) of the *Brettanomyces bruxellensis* life cycle.

Disassociate: Chemical process whereby a compound splits into ions.

Dolomite: A mineral and a sedimentary rock that is rich in magnesium and calcium carbonate.

Electron: Negatively charged subatomic particle that can be either bound to an atom or free.

Enzymes: Proteins that act as biological catalysts by speeding up chemical reactions.

***Epicoccum nigrum*:** Plant pathogen, a widespread fungus which produces coloured pigments that can be used as antifungal agents against other pathogenic fungi.

Fatty acids: Major component of fats that is used by plants and animals for energy and tissue development.

Feldspar: Group of rock-forming aluminium tectosilicate (three-dimensional tetrahedron structured silicate) minerals, also containing other cations such as sodium, calcium, potassium or barium.

Fungicide: Pesticide that kills or prevents the growth of fungi and their spores.

Glutamate: Naturally occurring compound that's found in many foods and gives them a unique flavour called umami.

Glycerol: Colourless, viscous liquid formed as a by-product in fermentation. It is used in industry as an emollient and laxative, and for making explosives and antifreeze.

Glycosidic bond: Covalent bond that connects a carbohydrate molecule, also known as a sugar, to another group.

Grafting: Horticultural technique whereby tissues of differing plants are joined so as to continue their growth together.

Heavy metal: Metal of relatively high density, or of high relative atomic weight.

Hippocampus: Complex brain structure embedded deep in the temporal lobe. It has a major role in learning and memory.

Humus: Dark, organic material that forms in soil when plant and animal matter decays.

Hydroponics: Type of horticulture which involves growing plants, usually crops or medicinal plants, without soil, by using water-based mineral nutrient solutions in an artificial environment.

Hydroxyl group: Structure consisting of an oxygen atom with two lone pairs bonded to a hydrogen atom. Hydroxyl groups readily participate in hydrogen bonding.

Ion: Atom or molecule with a net electrical charge.

Ionic lattice: A regular, repeating arrangement of ions.

Krebs cycle: Sequence of reactions by which most living cells generate energy during the process of aerobic respiration.

Lignin: Complex oxygen-containing organic polymer that, with cellulose, forms the chief constituent of wood.

Lipids: Group of organic compounds which include fats, waxes, sterols, fat-soluble vitamins and others. The functions of lipids include storing energy, signalling and acting as structural components of cell membranes.

Litmus test: Test to establish the acidity or alkalinity of a substance.

Loam: Soil with roughly equal proportions of sand, silt and clay.

Mafic: Silicate mineral or igneous rock rich in magnesium and iron. Most mafic minerals are dark in colour.

Maillard reaction: Chemical reaction between amino acids and sugars to create melanoidins. A process which gives browned food its distinctive flavour.

Master Sommelier/Master of Wine: Two qualifications that are regarded to be the highest standards of professional knowledge in the wine industry.

Mastication: Process of chewing food, the first step in digestion.

Metabolic reactions: Chemical processes that occur in living systems to alter molecules in order to release energy or convert small organic molecules into macromolecules.

Montmorillonite: Aluminium-rich clay mineral.

Mucilage: Thick, gluey substance produced by nearly all plants and some micro-organisms.

Mucosa: Soft tissue that lines the body's canals and organs in the digestive, respiratory and reproductive systems.

Mudstone: Dark, sedimentary rock formed from consolidated mud and lacking the laminations of shale.

Muscovite: An aluminium–potassium mineral, the most common member of the mica family.

Necrosis: Death of body tissue.

Neuron: Type of cell that receives and sends messages from the body to the brain and back to the body. A nerve cell.

New World: Wines produced outside the traditional wine-growing areas of Europe and the Middle East.

Nucleic acid: Large biomolecule that plays an essential role in all cells and viruses. A major function of nucleic acids involves the storage and expression of genomic information.

Nucleotide: Compound consisting of a nucleoside linked to a phosphate group. Nucleotides form the basic structural unit of nucleic acids such as DNA.

Nutrient antagonism: Overdoses of certain elements displacing other elements and preventing their uptake to a plant.

Oenology: The science and study of wine and winemaking.

Olfactory: Relating to the sense of smell.

Osmosis: Spontaneous passage or diffusion of water or other solvents through a semi-permeable membrane.

Pesticide: Substance used for destroying insects or other organisms harmful to cultivated plants or to animals.

Phloem: Living tissue in vascular plants that transports soluble organic compounds made during photosynthesis to the rest of the plant.

Photosynthesis: A process by which green plants and certain other organisms transform light energy into chemical energy.

***Phylloxera vastatrix*:** Almost microscopic, pale yellow, sap-sucking insect related to aphids, which feeds on the roots and leaves of grapevines.

Phytotoxicity: Term used to describe harmful effects that certain substances, such as chemicals or compounds, can have on plants.

PIWI: Fungus-resistant grape varieties that show particularly strong resistance to the vine diseases powdery and downy mildew.

Polymerize: Combine or cause to combine to form a polymer.

Polyphenols: Large family of naturally occurring phenols, abundant in plants and structurally diverse. Polyphenols include phenolic acids, flavonoids, tannic acid and ellagitannin, some of which have been used historically as dyes and for tanning garments.

Primary metabolites: Products directly involved in the growth, development and reproduction of living organisms.

Proton: Subatomic particle with a positive electrical charge. Protons are found in the atomic nucleus of every element.

Psychoactive: Substance that affects how the brain works and causes changes in mood, awareness, thoughts, feelings or behaviour.

PTSD: Short for post-traumatic stress disorder.

Pyrolysis: Thermal decomposition of materials at elevated temperatures.

Quartz: Hard mineral consisting of silica.

Quicklime: Product consisting chiefly of calcium oxide, obtained by roasting limestone, marble and shells. Uses include involvement in the production of steel, paper and glass, as well as within the construction industry for cement and mortar. In farming, it is included in Bordeaux mixture as the active fungicide ingredient.

Respiration: Chemical process used by organisms to release the energy from food, typically involving the consumption of oxygen and release of carbon dioxide.

Rhizodeposits: Material lost from plant roots, including water-soluble exudates and secretions of insoluble materials.

Rhizosphere: Region of soil or substrate that is directly influenced by root secretions and associated soil micro-organisms.

Root cortex: Outer layer of a plant root, located between the epidermis and the endodermis.

Secondary metabolite: Intermediate product of metabolism not vital for the growth and life of organisms but essential for the interaction of organisms with their environment, or produced in response to stress.

Shale: Soft, grey rock, usually formed from clay that has become hard.

Shorter chain acids: Organic acids that contain fewer than six carbon atoms in their carbon chain.

Sotolon: Major aroma and flavour component of fenugreek seed, lovage and artificial maple syrup.

Stomata: Tiny openings present on the epidermis of leaves. They play an important role in gaseous exchange and photosynthesis.

Terpenoid: Largest and structurally most diverse group of secondary metabolites derived from natural sources.

Terroir: French term encompassing the unique environmental factors shaping a wine's character. These factors include soil composition, climate and microclimate.

Thermal conductivity: Measure of a material's ability to conduct heat.

Thiamine: Vitamin B1; an essential micronutrient for humans and animals.

Tilling: Agricultural preparation of soil by mechanical agitation of various types, such as digging, stirring and overturning.

Transpiration: Physiological loss of water from plants in the form of water vapour, mainly from the stomata in leaves, but also through evaporation from the surfaces of leaves, flowers and stems.

Triumph Bonneville: An iconic motorcycle produced by Triumph Motorcycles, likened in style to the infamous TR6 Trophy driven by Steve McQueen in the 1963 film *The Great Escape*.

Vacuole: Space within a cell that is empty of cytoplasm (cellular fluid), lined with a membrane, and filled with fluid.

Veraison: Stage in the growing process when the grape begins to soften and change colour on the vine, indicating the onset of ripening.

Vigour: Amount of green growth, shoots, leaves and grapes produced by a grapevine over the course of a season. If a vine has many long shoots as well as larger leaves and yields a large crop, it has a lot of vigour.

***Vitis vinifera*:** The most common species of grapevine used in wine production.

Wernicke's area: Located in the left posterior superior temporal lobe, this area of the brain is responsible for understanding spoken and written language.

References

Chapter 1: Roots

D. A. Brown and William A. Albrecht (1951), 'Plant nutrition and the hydrogen ion. VII, Cation exchange between hydrogen clay and soils'. *Research Bulletin* (University of Missouri. Agricultural Experiment Station), 0477.

Bruce Dunn, Richard Leckie and Hardeep Singh (2017), 'Mycorrhizal Fungi'. Oklahoma State University Extension Id: HLA-6449.

Dave Greenshields (2013), 'Fungal contributions to the phosphorus cycle'. Joint Genome Institute project.

Monika Kozieł, Stefan Martyniuk and Grzegorz Siebielec (2021), 'Occurrence of *Azotobacter* spp. in cultivated soils in Poland'. *Polish Journal of Agronomy*, 44:3–8.

Michelle Ma (2020), 'Microbes help unlock phosphorus for plant growth'. *UW News* online https://www.washington.edu/news/2020/11/24/microbes-help-unlock-phosphorus-for-plant-growth/.

D. H. McNear Jr (2013), 'The rhizosphere – roots, soil and everything in between'. *Nature Education Knowledge* 4: article 1.

R. Nogueira de Sousa (2023), 'Introductory Chapter: Mycorrhizal Fungi – A Current Overview on Agricultural Productivity and Soil Health' in Nogueira de Sousa, R., ed., *Arbuscular Mycorrhizal Fungi in Agriculture – New Insights*. London: IntechOpen.

Alice Vacani (2017), 'The evolution of oenological chemistry in winemaking, antiquity to 20th century'. A thesis submitted for the Honour School of Chemistry, University of Oxford.

Yara Australia (2024), 'Nutrient deficiencies–Grapevines'. *Yara.com.au.*

Chapter 2: Debunking 'terroir'

Ben Anderson (2023), 'Soil Sulfur: How It Works'. *Epicgardening.com.*

Matthew Bush, Alison Smith and Eleni Vikeli (2017), 'Meet the Molecules; Geosmin'. John Innes Centre (www.jic.ac.uk/blog/meet-the-molecules-geosmin).

L. Chvyl, C. Williams and Mardi L. Longbottom (2010), 'Phosphorus fertilisation'. *Viti Notes [Grapevine Nutrition]*. Australian Wine Research Institute.

Jo Cowderoy (2022), 'The secret life of nitrogen'. *Vineyard Magazine.*

Jeannie Evers (2014), 'Weathering'. *National Geographic* (education.nationalgeographic.org/resource/weathering).

Lynn Fichter (2000), 'A Basic Sedimentary Rock Classification'. College of Science and Mathematics, James Madison University. csmgeo.csm.jmu.edu/geollab/fichter/sedrx/sedclass.html

Ted Goldammer (2021), *Grape Grower's Handbook: A Guide to Viticulture for Wine Production.* (Haymarket, VA: Apex Publishers).

Alfred E. Hartemink and Budiman Minasny (2014), 'Towards digital soil morphometrics'. *Geoderma*, 230–31:305–17.

E. M. Holbrook, C. A. Zambrano, C. T. O. Wright, et al. (2023), '*Mycobacterium vaccae* NCTC 11659, a soil-derived bacterium with stress resilience properties, modulates the proinflammatory effects of LPS in macrophages'. *International Journal of Molecular Sciences*, 24(6):5176.

Ronald S. Jackson (2020), 'Site selection and climate' in *Wine Science* (Elsevier, Academic Press), 331–74.

Hobart M. King (n.d.), 'Pumice'. *Geology.com.* https://geology.com/rocks/pumice.shtml

Andrei Kuzin and Alexei Solovchenko (2021), 'Essential role of potassium in apple and its implications for Management of orchard fertilization'. *Plants*, 10(12):2624.

Sarah Langford (2023), *Rooted: How Regenerative Farming Can Change the World* (Penguin Random House).

Alex Maltman (2018), *Vineyards, Rocks, and Soils: The Wine Lover's Guide to Geology.* (New York: Oxford Academic).

L. A. Morris (2004), 'Soil biology and tree growth: soil organic matter forms and functions' in Evans, J. and Youngquist, J. A., *Encyclopedia of Forest Sciences* (Elsevier, Academic Press), 1201–7.

M. E. R. O'Brien, H. Anderson, E. Kaukel, et al. (2004), 'SRL172 (killed *Mycobacterium vaccae*) in addition to standard chemotherapy improves quality of life without affecting survival, in patients with advanced non-small-cell lung cancer: phase III results'. *Annals of Oncology*, 15(6):906–14.

Ruth Rhodes, Neil Miles and Jeffrey Charles Hughes (2018), 'Interactions between potassium, calcium and magnesium in sugarcane grown on two contrasting soils in South Africa'. *Field Crops Research*, 225:1–11.

The editors of the Encyclopaedia (2024), 'humus'. *Encyclopedia Britannica* online https://www.britannica.com/science/humus-soil-component

Chapter 3: Natural law

T. R. Aguiar, F. R. Bortolozo, F. A. Hansel, et al. (2015), 'Riparian buffer zones as pesticide filters of no-till crops'. *Environmental Science and Pollution Research*, 22:10618–26.

Mahdi Al-Kaisi, Mark Hanna and Michael Tidman (2004), 'Frequent tillage and its impact on soil quality'. *IC* 492(13)78–79. Iowa State University – Midwest Grape and Wine Industry Institute.

Biodynamic Federation – Demeter International e.V. (2022), 'Production, Processing and Labelling – International Standard for the use and certification of Demeter, Biodynamic and related trademarks (as of: Oct 2022)'.

M. Bonato, E. A. Martin, A. F. Cord, et al. (2023), 'Applying generic landscape-scale models of natural pest control to real data: associations between crops, pests and biocontrol agents make the difference'. *Agriculture, Ecosystems & Environment*, 342:108215.

Thomas Brown (2023), 'Agricultural fungicides: impact on long-term food and biological security'. House of Lords Library: https://lordslibrary.parliament.uk/agricultural-fungicides-impact-on-long-term-food-and-biological-security/

F. C. Coelho, R. Squitti, M. Ventriglia, et al. (2020), 'Agricultural use of copper and its link to Alzheimer's disease'. *Biomolecules* 10(6):897, https://eur-lex.europa.eu/eli/reg_impl/2011/540/oj

Department for Environment, Food and Rural Affairs (13 June 2022), *Government food strategy*. CP 698.

Jean-Michel Florin (2021), *Biodynamic Wine Growing: Understanding the Vine and Its Rhythms* (Edinburgh: Floris Books).

House of Commons Environment, Food and Rural Affairs Committee (2023), *Soil Health: First Report of Session 2023–24*. https://committees.parliament.uk/publications/42415/documents/210844/default/

IFOAM EU Group (2013), *EU rules for organic wine production*. IFOAM EU.

Britt and Per Karlsson (2014), *Biodynamic, Organic and Natural Winemaking: Sustainable Viticulture and Viniculture* (Edinburgh: Floris Books).

Lord Northbourne (1940), *Look to the Land* (London: Dent).

John Paull (2019), 'The pioneers of biodynamics in Great Britain: from anthroposophic farming to organic agriculture (1924–1940)'. *Journal of Environment Protection and Sustainable Development*, 5(4):138–45.

Rebecca Robertson and Jonathan Wentworth (2020), 'Understanding insect decline: data and drivers'. The Parliamentary Office of Science and Technology (POST).

R. Schulz, S. Bub, L. L. Petschick, et al. (2021), 'Applied pesticide toxicity shifts toward plants and invertebrates, even in GM Crops'. *Science*, 372(6537):81–4.

V. Silva, H. G. J. Mol, P. Zomer, et al. (2019), 'Pesticide residues in European agricultural soils a hidden reality unfolded'. *Science of the Total Environment*, 653:1532–45.

Peter Staudenmaier (2013), 'Organic Farming in Nazi Germany: the politics of biodynamic agriculture, 1933–1945. *Environmental History*, 18:383–411.

Rudolf Steiner (2001), *Colour: Cw 291* (Forest Row, East Sussex: Rudolf Steiner Press).

Dietrich Stoltzenberg (2004), *Fritz Haber: Chemist, Nobel Laureate, German, Jew* (Chemical Heritage Foundation).

R. Teysseire, G. Manangama, I. Baldi, et al. (2021), 'Determinants of non-dietary exposure to agricultural pesticides in populations living close to fields: a systematic review'. *Science of the Total Environment*, 761:143294.

Chapter 4: The anatomy of a grape

M. S. Allen, M. J. Lacey, R. L. N. Harris and W. V. Brown (1991), 'Contribution of methoxypyrazines to Sauvignon Blanc wine aroma'. *American Journal of Enology and Viticulture*, 42:109–12.

Jean-François Arnoldi, Sylvain Coq, Sonia Kéfi and Sébastien Ibanez (2019), 'Positive plant–soil feedback trigger tannin evolution by niche construction: a spatial stoichiometric model'. *Journal of Ecology*, 108(1):378–91.

N. Bandara and M. Chalamaiah (2019), 'Bioactives from agricultural processing py-products' in Melton, L., Shahidi, F. and Varelis, P., eds, *Encyclopedia of Food Chemistry* (Elsevier).

Jacqueline Barona, Juan C. Aristizabal, Christopher N. Blesso, et al. (2012), 'Grape polyphenols reduce blood pressure and increase flow-mediated vasodilation in men with metabolic syndrome'. *The Journal of Nutrition*, 142(9):1626–32.

C. E. Burtch, A. K. Mansfield and D. C. Manns (2017), 'Reaction kinetics of monomeric anthocyanin conversion to polymeric pigments and their significance to color in interspecific hybrid wines'. *Journal of Agricultural and Food Chemistry*, 65(31):6379–86.

Murli Dharmadhikari (2021), 'Composition of Grapes'. Iowa State University – Midwest Grape and Wine Industry Institute.

Natacha Fontes, Hernâni Gerós and Serge Delrot (2011), 'Grape berry vacuole: a complex and heterogeneous membrane system specialized in the accumulation of solutes'. *American Journal of Enology and Viticulture*, 62:270–8.

Olivier Geffroy, Didier Kleiber and Alban Jacques (2020), 'Key facts about rotundone and practical ways to pepper your wine with this fascinating aroma compound'. *IVES Technical Reviews: Vine and Wine* (September).

Jamie Goode (2014), *Wine Science: The Application of Science in Winemaking*, 2nd edition (London: Mitchell Beazley).

Dylan Grigg (2017), 'An investigation into the effect of grapevine age on vine performance, grape and wine composition, sensory evaluation and epigenetic characterisation'. Thesis submitted to the School of Agriculture, Food and Wine of the University of Adelaide.

A. D. Harner, J. E. Vanden Heuvel, R. P. Marini, et al. (2019), 'Modeling the impacts of weather and cultural factors on rotundone concentration in cool-climate Noiret wine grapes'. *Frontiers in Plant Science*, 10:1255.

Ádám István Hegyi, Margot Otto, József Geml, et al. (2023), 'The origin of the particular aroma of noble rot wines: various fungi contribute to the development of the aroma profile of botrytised grape berries'. *OENO One*, 57(3):165–76.

H. E. Khoo, A. Azlan, S. T. Tang and S. M. Lim (2017), 'Anthocyanidins and anthocyanins: colored pigments as food, pharmaceutical ingredients, and the potential health benefits'. *Food & Nutrition Research*, 61(1):1361779.

B. S. Mpelasoka, D. P. Schachtman, M. T. Treeby and M. R. Thomas (2003), 'A review of potassium nutrition in grapevines with special emphasis on berry accumulation'. *Australian Journal of Grape and Wine Research*, 9:154–68.

C. Müller and M. Riederer (2005), 'Plant surface properties in chemical ecology'. *Journal of Chemical Ecology*, 31(11):2621–51.

Alfredo S. Negri, Bhakti Prinsi, Osvaldo Failla, et al. (2015), 'Proteomic and metabolic traits of grape exocarp to explain different anthocyanin concentrations of the cultivars'. *Frontiers in Plant Science*, 6: article 603.

A. N. Panche, A. D. Diwan and S. R. Chandra (2016), 'Flavonoids: an overview'. *Journal of Nutritional Science*, 5:e47.

Steven F. Price (1994), 'Sun exposure and flavanols in grapes'. A Thesis Submitted to Oregon State University.

B. C. Rankine and K. F. Pocock (1969), 'β-Phenethanol and η-hexanol in wines: influence of yeast strain, grape variety and other factors; and taste thresholds'. *Vitis: Journal of Grapevine Research*, 8(1):23–37.

A. M. S. Riel, N. B. Wageling, D. A. Decato and O. B. Berryman (2017), 'Anion–arene interactions and the anion–π phenomenon' in Atwood J. L., ed., *Comprehensive Supramolecular Chemistry II* (Elsevier).

V. Riffle, N. Palmer, L. F. Casassa and J. C. Dodson Peterson (2021), 'The effect of grapevine age (*Vitis vinifera* L. cv. Zinfandel) on phenology and gas exchange parameters over consecutive growing seasons'. *Plants,* 10(2):311.

Margit Rid, Anna Markheiser, Christoph Hoffmann and Jürgen Gross (2018), 'Waxy bloom on grape berry surface is one important factor for oviposition of European grapevine moths'. *Journal of Pest Science,* 91:1225–39.

Anthony L. Robinson, Paul K. Boss, Peter S. Solomon, et al. (2014), 'Origins of grape and wine aroma. Part 1. Chemical components and viticultural impacts'. *The American Journal of Enology and Viticulture,* 65:1.

Sascha Rohn (2019), 'Covalent interactions between proteins and phenolic compounds' in Melton, L., Shahidi, F. and Varelis, P., eds, *Encyclopedia of Food Chemistry* (Elsevier).

Laura A. de la Rosa, Jesús Omar Moreno-Escamilla, Joaquín Rodrigo-García and Emilio Alvarez-Parrilla (2019), 'Phenolic compounds' in Yahia, E. M., ed., *Postharvest Physiology and Biochemistry of Fruits and Vegetables* (Elsevier, Woodhead Publishing).

V. Sichel, G. Sarah and N. Girollet (2023), 'Chimeras in Merlot grapevine revealed by phased assembly'. *BMC Genomics,* 24:396.

Amelie Slegers, Paul Angers, Etienne Ouellet, et al. (2015), 'Volatile compounds from grape skin, juice and wine from five interspecific hybrid grape cultivars grown in Quebec (Canada) for wine production'. *Molecules,* 20(6):10980–1016.

S. Vezzulli, L. Leonardelli, U. Malossini, et al. (2012), 'Pinot Blanc and Pinot Gris arose as independent somatic mutations of Pinot Noir'. *Journal of Experimental Botany,* 63(18):6359–69.

Leroy G. Wade (2018), 'phenol'. *Encyclopedia Britannica* online https://www.britannica.com/science/phenol.

R. P. Walker, C. Bonghi, S. Varotto, et al. (2021), 'Sucrose metabolism and transport in grapevines, with emphasis on berries and leaves and insights gained from a cross-species comparison'. *International Journal of Molecular Sciences,* 22(15): article 7794.

Andrew L. Waterhouse, Gavin L. Sacks and David W. Jeffery (2016), 'Methoxypyrazine' in *Understanding Wine Chemistry* (Wiley).

Chapter 5: Microbes

E. Aslankoohi, M. N. Rezaei, Y. Vervoort, et al. (2015), 'Glycerol production by fermenting yeast cells is essential for optimal bread dough fermentation'. *PloS ONE,* 10(3):e0119364.

M. Baron, B. Prusova, L. Tomaskova, et al. (2017), 'Terpene content of wine from the aromatic grape variety "Irsai Oliver" (*Vitis vinifera* L.) depends on maceration time'. *Open Life Sciences,* 12(1):42–50.

R. Baumes, J. Wirth, S. Bureau, et al. (2002), 'Biogeneration of C 13-norisoprenoid compounds: experiments supportive for an apo-carotenoid pathway in grapevines'. *Analytica Chimica Acta,* 458(1):3–14.

K. A. Bindon, P. R. Dry and B. R. Loveys (2007), 'Influence of plant water status on the production of C13-norisoprenoid precursors in *Vitis vinifera* L. cv. Cabernet Sauvignon grape berries'. *Journal of Agricultural and Food Chemistry*, 55:4493–500.

P. Commenil, L. Belingheri, G. Bauw and B. Dehorter (1999), 'Molecular characterization of a lipase induced in *Botrytis cinerea* by components of grape berry cuticle'. *Physiological and Molecular Plant Pathology*, 55(1):37–43.

Murli Dharmadhikari (2021), 'Yeast Autolysis'. Iowa State University – Midwest Grape and Wine Industry Institute.

Alessandro Genovese, Paola Piombino, Maria Tiziana Lisanti and Luigi Moio (2005), 'Occurrence of furaneol (4-hydroxy-2,5-dimethyl-3(2H)-furanone) in some wines from Italian native grapes'. *Annali di Chimica*, 95(6): 415–19

Harold Mcgee (2020), *Nose Dive: A Field Guide to the World's Smells* (London: John Murray).

S. Maicas (2020), 'The role of yeasts in fermentation processes'. *Microorganisms*, 8(8):1142.

T. Maoka (2020), 'Carotenoids as natural functional pigments'. *Journal of Natural Medicines*, 74(1):1–16.

V. Martin, M. J. Valera, K. Medina, et al. (2018), 'Oenological impact of the *Hanseniaspora/Kloeckera* yeast genus on wines—a review'. *Fermentation*, 4(3):76.

M. Mendes-Pinto (2009), 'Carotenoid breakdown products the—norisoprenoids—in wine aroma'. *Archives of Biochemistry and Biophysics*, 483(2):236–45.

P. Nicolle, A. Gerzhova, A. Roland, et al. (2022), 'Thiol precursors and amino acids profile of white interspecific hybrid *Vitis* varieties and impact of foliar urea and sulfur supplementation on the concentration of thiol precursors in *Vitis* sp. Vidal berries'. *OENO One*, 56(3):357–69.

G. L. Sacks, M. J. Gates, F. X. Ferry, et al. (2012), 'Sensory threshold of 1,1,6-trimethyl-1,2-dihydronaphthalene (TDN) and concentrations in young Riesling and non-Riesling wines'. *Journal of Agricultural and Food Chemistry*, 60(12):2998–3004.

M. A. Sefton, G. K. Skouroumounis, G. M. Elsey and D. K. Taylor (2011), 'Occurrence, sensory impact, formation, and fate of damascenone in grapes, wines, and other foods and beverages'. *Journal of Agricultural and Food Chemistry* 59(18):9717–46.

José Sousa Câmara, José C. Marques, Maria Arminda Alves and António C. Silva Ferreira (2004), '3-Hydroxy-4,5-dimethyl-2(5H)-furanone levels in fortified Madeira wines: relationship to sugar content'. *Journal of Agricultural and Food Chemistry*, 52(22):6765–9.

J. H. Swiegers, E. J. Bartowsky, P. A. Henschke and I. Pretorius (2005), 'Yeast and bacterial modulation of wine aroma and flavour'. *Australian Journal of Grape and Wine Research*, 11(2):139–73.

Chris van Tulleken (2023), *Ultra-Processed People: Why Do We All Eat Stuff That Isn't Food ... and Why Can't We Stop?* (London: Cornerstone Press).

Kelli White (2018), '*Brettanomyces*: Science and Context'. GuildSomm.

L. Zhu, M. Zhang, Z. Liu, et al. (2019), 'Levels of furaneol in Msalais wines: a comprehensive overview of multiple stages and pathways of its formation during Msalais winemaking'. *Molecules*, 24(17):3104.

D. E. Zohre and H. Erten (2002), 'The influence of *Kloeckera apiculata* and *Candida pulcherrima* yeasts on wine fermentation'. *Process Biochemistry*, 38(3):319–24.

Chapter 6: Manipulations at the winery

Veronica Avram, Calin Floare, Anamaria Hosu, et al. (2015), 'Characterization of Romanian wines by gas chromatography–mass spectrometry'. *Analytical Letters,* 48(7):1099–116.

R. B. Boulton, V. L. Singleston, L. F. Bisson and R. E. Kunkee (1999), 'The role of sulfur dioxide in wine'. *Principles and Practices of Winemaking* (New York: Springer).

E. Cadahía, S. Varea, L. Muñoz, et al. (2001), 'Evolution of ellagitannins in Spanish, French, and American oak woods during natural seasoning and toasting'. *Journal of Agricultural and Food Chemistry,* 49(8):3677–84.

P. Chatonnet and D. Dubourdieu (1998), 'Comparative study of the characteristics of American white oak (*Quercus alba*) and European oak (*Quercus petraea* and *Q. robur*) for production of barrels used in barrel aging of wines'. *American Journal of Enology and Viticulture,* 49:79–85.

V. Ducruet (1984), 'Comparison of the headspace volatiles of carbonic maceration and traditional wine'. *Lebensmittel Wissenschaft & Technologie,* 17(4):217–21.

J. Echave, M. Barral, M. Fraga-Corral, et al. (2021), 'Bottle aging and storage of wines: a review'. *Molecules,* 26(3):713.

I. Etaio, F. J. P. Elortondo, M. Albisu, et al. (2008), 'Effect of winemaking process and addition of white grapes on the sensory and physicochemical characteristics of young red wines'. *Australian Journal of Grape and Wine Research,* 14:211–22.

V. Ferreira and R. Lopez (2019), 'The actual and potential aroma of winemaking grapes'. *Biomolecules,* 9(12):818.

C. Flanzy, M. Flanzy and P. Benard (1987), *La Vinification par Macération Carbonique.* Paris: INRA.

R. Gawel, M. Day, S. C. Van Sluyter, et al. (2014), 'White wine taste and mouthfeel as affected by juice extraction and processing'. *Journal of Agricultural and Food Chemistry,* 62(41):10008–14.

Pat Henderson (2009), 'Sulfur dioxide: science behind this anti-microbial, anti-oxidant, wine additive'. *Practical Winery & Vineyard Journal,* Jan/Feb.

D. Hernanz, A. F. Recamales, M. L. Gonzalez-Miret, et al. (2007), 'Phenolic composition of white wines with a prefermentative maceration at experimental and industrial scale'. *Journal of Food Engineering,* 80(1):327–35.

M. Navarro, N. Kontoudakis, T. Giordanengo, et al. (2016), 'Oxygen consumption by oak chips in a model wine solution: influence of the botanical origin, toast level and ellagitannin content'. *Food Chemistry,* 199:822–7.

Kurt Nordstrom (1962), 'Formation of ethyl acetate in fermentation with Brewer's yeast. II. Kinetics of formation from ethanol and influence of acetaldehyde'. *Journal of the Institure of Brewing,* 68(2):188–96.

Daniel Pambianchi (n.d.), 'Impact of barrel kinetics and dynamics on wine'. *WineMaker* online magazine.

V. Schneider (1998), 'Must hyperoxidation: a review'. *American Journal of Enology and Viticulture,* 49(1):65–73.

Beth Sissons (medical reviewer Philip Ngo) (2023), 'Why wine hangovers happen and how to treat them'. *Medical News Today* online.

T. N. Sneyd (1989), 'Carbonic maceration: an overview'. *Australian and New Zealand Wine Industry Journal,* 4(4):281–2.

Creina Stockley, Angelika Paschke-Kratzin, Pierre-Louis Teissedre, et al. (2021), *SO_2 and Wine: A Review,* 1st edn. Paris: OIV publications.

G. Styger, B. Prior and F. F. Bauer (2011), 'Wine flavor and aroma'. *Journal of Industrial Microbiology and Biotechnology* 38(9):1145.

Kaushlendra Tingi (2013), 'Pyrolysis kinetics of physical components of wood and wood-polymers using isoconversion method'. *Journal of Agriculture*, 3:12–32.

H. B. Wedler, R. P. Pemberton and D. J. Tantillo (2015), 'Carbocations and the Complex Flavor and Bouquet of Wine: Mechanistic Aspects of Terpene Biosynthesis in Wine Grapes'. *Molecules* 20(6):10781–92.

J. T. Weld and A. Gunther (1947), 'The Antibacterial Properties of Sulfur'. *Journal of Experimental Medicine* 85(5):531–42.

Chapter 7: Flavour perception

Alain Berthoz, trans. Giselle Weiss (2017), *The Vicarious Brain: Creator of Worlds.* (Cambridge, MA: Harvard University Press).

M. A. Buzalaf, A. R. Hannas and M. T. Kato (2012), 'Saliva and dental erosion'. *Journal of Applied Oral Science*, 20(5):493–502.

Peter Coucquyt, Bernard Lahousse and Johan Langenbick (2020), *The Art & Science of Foodpairing: 10,000 flavour matches that will transform the way you eat* (Richmond Hill, ON: Firefly Books).

U. Dicke and G. Roth (2016), 'Neuronal factors determining high intelligence'. *Philosophical Transactions of the Royal Society B: Biological Sciences*, 371(1685), 20150180.

Emma L. Feeney, Lauren McGuinness, John E. Hayes and Alissa A. Nolden (2021), 'Genetic variation in sensation affects food liking and intake'. *Current Opinion in Food Science*, 42:203–14.

Daniel Hwang (n.d.), 'Are you a super-taster?'. Online article, Institute for Molecular Bioscience, The University of Queensland, Australia.

M. D. Kaplan and B. J. Baum (1993), 'The functions of saliva'. *Dysphagia*, 8(3):225–9.

G. Morrot, F. Brochet and D. Dubourdieu (2001), 'The color of odors'. *Brain and Language*, 79(2):309–20.

Christopher P. Nichols, Julian A. Drewe, Robin Gill, et al. (2016), 'A novel causal mechanism for grey squirrel bark stripping: the calcium hypothesis'. *Forest Ecology and Management*, 367:12–20.

A. J. Scott-Thomas, M. Syhre, P. K. Pattemore, et al. (2010), '2-Aminoacetophenone as a potential breath biomarker for *Pseudomonas aeruginosa* in the cystic fibrosis lung'. *BMC Pulmonary Medicine*, 10:56.

Nik Sharma (2020), *The Flavor Equation: The Science of Great Cooking Explained.* (San Francisco: Chronicle Books).

Gordon M. Shepard (2016), *Neuroenology: How the Brain Creates the Taste of Wine.* (New York: Columbia University Press).

C. Spence and J. Youssef (2021), 'Aging and the (chemical) senses: implications for food behaviour amongst elderly consumers'. *Foods*, 10(1):168.

S. P. Wooding, V. A. Ramirez and M. Behrens (2021), 'Bitter taste receptors: genes, evolution and health'. *Evolution, Medicine, and Public Health*, 9(1):431–47.

Carl Zimmer (2013), 'The Smell of Evolution'. *National Geographic* online. https://www.nationalgeographic.com/science/article/the-smell-of-evolution

Index

Also published by Académie du Vin Library

Académie du Vin Library was founded by Steven Spurrier and friends, dedicated to publishing the finest wine writing, and it has grown into the world's leading wine book publisher in six short years. We choose our books with care – above all for their readability, but also because we genuinely believe they have something important to say about the world of fine wine that will enhance your drinking pleasure.

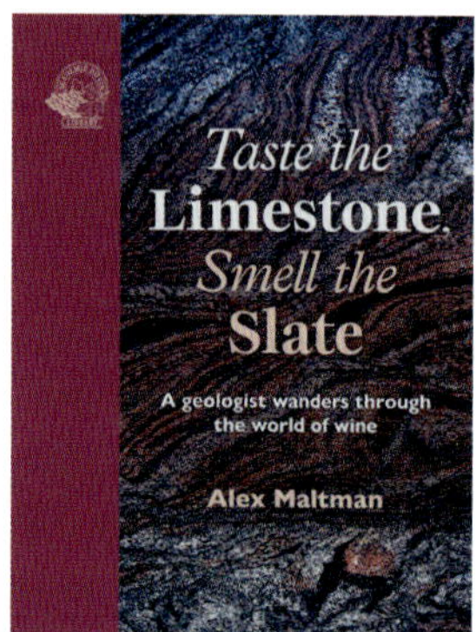

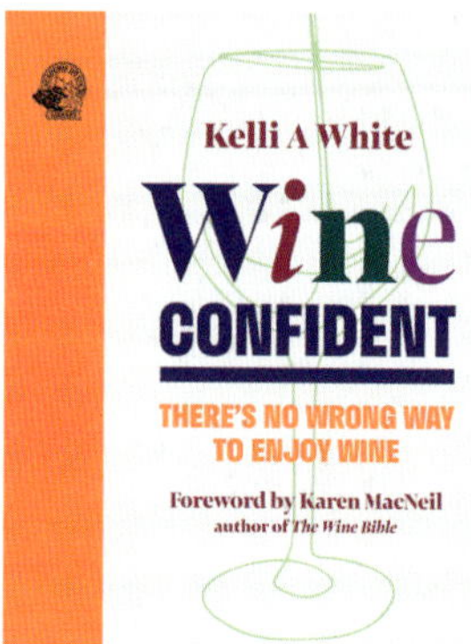

TASTE THE LIMESTONE, SMELL THE SLATE
Professor Alex Maltman
Alex Maltman, emeritus professor of earth sciences at Aberystwyth University, finds himself between a rock and a vineyard place as he explains how a wine's flavours relate to the geology at foot, and discovers that there is more to 'minerality' than mud, rocks and the earth's stark materials.

WINE CONFIDENT: There's No Wrong Way to Enjoy Wine
Kelli A White
Kelli White loves every aspect of wine and wants to ignite the same kind of passion in her readers. For those who have fallen in love with wine already, this book is an imaginative and practical guide to embarking on its greatest adventures. For those who haven't yet, it is the spark to light the fuse.

BEHIND THE GLASS: The Chemical and Sensorial Terroir of Wine Tasting
Gus Zhu MW
What makes red wine red? Do genetic differences, culture and life experience change our perception of wine? And is there science behind the obscure language of tasting notes? The answers to all of these questions and more are explored in *Behind the Glass*, a smart, accessible investigation into the science behind a glass of wine – and our appreciation of it.

Browse the full list and buy online at academieduvinlibrary.com